REGIONAL PHYSICAL MAPPING

SERIES EDITORS
Kay E. Davies, *University of Oxford*
Shirley M. Tilghman, *Princeton University*

The identification and mapping of genes, analysis of their structures, and discovery of the functions they encode are now cornerstones of experimental biology, health research, and biotechnology. *Genome Analysis* is a series of short, single-theme books that review the data, methods, and ideas emerging from the study of genetic information in humans and other species. Each volume contains invited papers that are timely, informative, and concise. These books are an information source for junior and senior investigators in all branches of biomedicine interested in this new and fruitful field of research.

SERIES VOLUMES
1. Genetic and Physical Mapping
2. Gene Expression and Its Control
3. Genes and Phenotypes
4. Strategies for Physical Mapping
5. Regional Physical Mapping

Forthcoming

6. Genome Maps and Neurological Disorders
7. Genome Rearrangement and Stability

REGIONAL PHYSICAL MAPPING

Edited by
Kay E. Davies
University of Oxford

Shirley M. Tilghman
Princeton University

Volume 5 / GENOME ANALYSIS

 Cold Spring Harbor Laboratory Press 1993

Genome Analysis Volume 5
Regional Physical Mapping

All rights reserved
Copyright 1993 by Cold Spring Harbor Laboratory Press
Printed in the United States of America
ISBN 0-87969-413-0
ISSN 1050-8430
LC 93-70644

Cover and book design by Leon Bolognesi & Associates, Inc.

Authorization to photocopy items for internal or personal use, or the internal or personal use of specific clients, is granted by Cold Spring Harbor Laboratory Press for libraries and other users registered with the Copyright Clearance Center (CCC) Transactional Reporting Service, provided that the base fee of $5.00 per article is paid directly to CCC, 27 Congress St., Salem, MA 01970. [0-87969-413-0/93 $5.00 + .00]. This consent does not extend to other kinds of copying, such as copying for general distribution, for advertising or promotional purposes, for creating new collective works, or for resale.

All Cold Spring Harbor Laboratory Press publications may be ordered directly from Cold Spring Harbor Laboratory Press, 10 Skyline Drive, Plainview, New York 11803-2500. Phone: 1-800-843-4388 in Continental U.S. and Canada. All other locations (516) 349-1930.

Contents

Preface *vii*

The Human Major Histocompatibility Complex: A 4000-kb Segment of the Human Genome Replete with Genes **1**
R. Duncan Campbell

Molecular Genetics and Physical Mapping in Human Xp21 **35**
Anthony P. Monaco, Ann P. Walker, Meng F. Ho, Jamel Chelly, Edward Clarke, Yumiko Ishikawa-Brush, and Françoise Muscatelli

Large-scale DNA Sequence Analyses of Mammalian T-Cell Receptor Loci **63**
Leroy Hood, Ben F. Koop, Lee Rowen, and Kai Wang

Construction of a 10-Mb Physical Map in the Adenomatous Polyposis Region of Chromosome 5 **89**
Ellen Solomon and Anna-Maria Frischauf

Structure of the Terminal Region of the Short Arm of Chromosome 16 **107**
Peter C. Harris and Douglas R. Higgs

Index *135*

Preface

Developments in genome mapping have continued at a rapid pace since the last volume of this series was published. In particular, YAC contig maps of whole human chromosomes have been constructed, and a high-resolution genetic map based on microsatellite markers has been defined. These are landmarks in human genome mapping because they indicate that the scientific community is now very close to having a complete physical map integrated with a genetic map. Armed with these tools, the molecular biologist can move forward into the detailed analysis of genes, genome structure, and sequence.

The chapters in this volume describe the detailed analysis of several regions of the human genome that have already been extensively characterized. Each one tells a different story: Duncan Campbell describes the remarkable gene density found at the MHC locus and, at the other extreme, Tony Monaco and his co-workers present the structure at Xp21, where there are only a few genes but where the genes are exceptionally large. Peter Harris and Doug Higgs describe the study of the telomeric region of 16p, which reveals not only a gene-dense region, but also an interesting genomic structure that displays long-range polymorphism. The problems of constructing long YAC contigs around tumor suppressor genes are discussed by Ellen Solomon and Anna-Maria Frischauf. Finally, Lee Hood and his colleagues summarize their findings in the initial stages of the large-scale sequencing of the mammalian T-cell receptor loci. The arguments for total genomic sequencing are presented. The comparative sequencing presented in this chapter is discussed, together with the way in which this information may be used to solve the problems of gene function. As sequencing and mapping technologies become more efficient, this volume provides a glimpse of the fascinating biology that will be revealed through the human genome mapping project.

We are grateful to all of the authors in this volume for their hard work in presenting such excellent contributions. We are also grateful to the staff of the Cold Spring Harbor Laboratory Press, especially Nancy

Ford and her colleagues Patricia Barker and Mary Cozza, whose enthusiasm and determination ensure that the volumes appear as rapidly as possible.

Kay E. Davies
Shirley M. Tilghman
February 1993

REGIONAL PHYSICAL MAPPING

The Human Major Histocompatibility Complex: A 4000-kb Segment of the Human Genome Replete with Genes

R. Duncan Campbell

MRC Immunochemistry Unit
Department of Biochemistry
Oxford OX1 3QU, United Kingdom

The human major histocompatibility complex (MHC), which is located in the short arm of chromosome 6 in the 6p21.3 band, has attracted considerable attention because it contains genes encoding cell-surface molecules that are essential for mounting an immune response to pathogenic organisms. In recent years, the MHC has been characterized in detail by a combination of physical mapping and molecular cloning. Pulsed field gel electrophoresis (PFGE) has been used to establish linkage of genes to the MHC, and a long-range restriction map of this chromosomal segment extending over 4000 kb of DNA has been constructed. The whole of the MHC has now been cloned in cosmid and/or yeast artificial chromosome (YAC) vectors, and large portions have been characterized for the presence of genes. This work has revealed that the MHC contains at least 80 genes, both housekeeping and tissue-specific, with the gene density in some regions approaching one gene every few kilobases. Some of the genes encode proteins involved in different aspects of the immune response, whereas others encode proteins with no obvious role in immunity. The finding of such a large number of genes within the MHC strengthens the notion that there is a nonrandom distribution of genes in the human genome.

Main points discussed include:

❏ a brief historical account of the development of a genetic map of the MHC

❏ the use of PFGE to construct detailed physical maps of the human MHC

❏ the cloning of the human MHC and the techniques used to define the large number of novel genes that have now been located in the human MHC

❏ a brief account of what the functions of some of the novel gene products appear to be

INTRODUCTION

The MHC is a cluster of loci coding for polymorphic proteins involved in the presentation of antigen to thymus-derived (T) lymphocytes. This definition refers specifically to the class I and class II loci, since historically it was the study of the products of these genes that led to the discovery of this genetic complex. However, the term MHC is also used here in reference to the chromosomal segment that contains not only the HLA class I and class II genes, but also the large number of unrelated genes which have now been located within the MHC.

The human MHC has been mapped to the short arm of chromosome 6 in the distal portion of the 6p21.3 band (Franke and Pellegrino 1977; Spring et al. 1985), whereas the murine MHC, termed the H-2 complex, is located on chromosome 17. The human MHC can be conveniently divided into three major regions (Fig. 1). The class I and class II regions each encode highly polymorphic families of cell-surface glycoproteins that are involved in the presentation of antigenic peptides to T lymphocytes during an immune response (Davis and Bjorkman 1988; Townsend and Bodmer 1989; Bjorkman and Parham 1990; Braciale and Braciale 1991; van Bleek and Nathenson 1992). The class I region contains at least 18 highly related genes which include those encoding the classical transplantation antigens HLA-A, -B, and -C. The class II region (HLA-D) is arranged into subregions DP, DQ, and DR, each containing at least one A and B pair of genes encoding the α and β polypeptide chains of the class II molecule. A number of other class-II-related sequences (including DO, DN, and DM) have also been defined within the HLA-D region. The class I and class II regions are separated by the class III region, which also contains genes that encode proteins with immune-related functions such as the complement components C2, C4, and Factor B, the cytokines tumor necrosis factors (TNF) A and B, and three members of the major heat shock protein HSP70 family. The

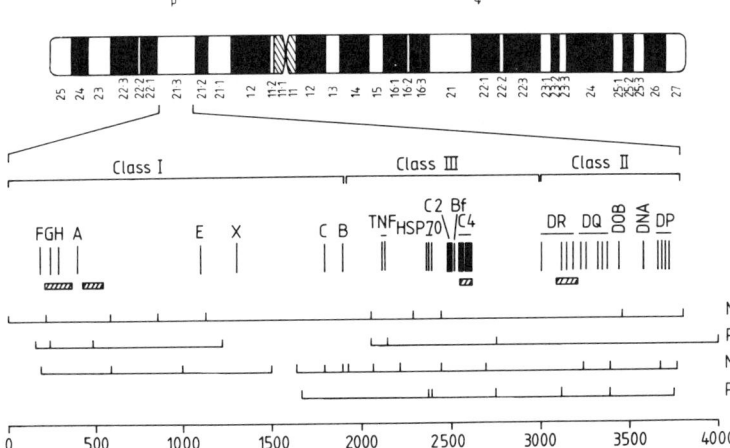

Figure 1 Schematic representation of human chromosome 6 (*top*) with the Giemsa-light and -dark bands shown in white and black, respectively. The chromosomal segment in the 6p21.3 band corresponding to the MHC is shown enlarged in the middle of the figure. The extents of the class I, class II, and class III regions are marked; the genes are represented by bars. Those regions which have been shown to vary between haplotypes are indicated by cross-hatched bars. A long-range restriction map of the MHC constructed from data derived from PFGE studies (see text for references) is shown at the bottom of the figure. Enzyme sites shown are for N, *Not*I; R, *Nru*I; M, *Mlu*I; P, *Pvu*I.

MHC has been studied most extensively in the mouse and in man, and it is initially worth considering the historical development of our present understanding of the MHCs in these two species.

GENETIC MAP

It has long been known that tumors from a particular mouse only grow in mice of the same inbred strain. Genetic analysis of tumor rejection between inbred strains demonstrated that one of the loci responsible was identical to a gene controlling the presence of a blood group antigen (Gorer 1937). Subsequently, the genetic organization of the loci encoding the cell-surface structures responsible for graft rejection was extensively characterized utilizing alloantisera prepared by cross-immunization between congenic, and later recombinant congenic, mice. This work revealed that the mouse H-2 (histocompatibility-2) locus was a complex of two genetically separable genes or gene clusters (H-2K and H-2D) each encoding H-2 antigens. At each locus, multiple allelic variants were possible, giving rise to the complex serology observed with alloantisera (for review, see Klein 1975). In addition, genetic control of a quantitative variant in the serum of normal mice (serum substance, Ss) was shown to be encoded by a locus lying within the H-2 complex (Shreffler 1964). Serum substance was later identified as the mouse complement component C4 (Lachmann et al. 1975; Meo et al. 1975).

Unrelated studies on the antibody response of inbred mouse strains to synthetic branched polypeptides showed a high degree of polymorphism in the ability of different mouse strains to mount a response to antigenic challenge. It was demonstrated that this immune responsiveness was under genetic control and that the "immune response" (Ir) genes were very closely linked to the H-2 complex (McDevitt and Chintz 1969). McDevitt et al. (1972) then mapped the Ir gene controlling the response to a specific polypeptide to a distinct region, the I region, lying between the H-2K and Ss loci. Thus, the classical genetic studies defined the existence of a cluster of loci lying on chromosome 17 of the mouse (Klein and Figuero 1986) that display remarkable polymorphism and control both transplantation antigens and the immune response (for review, see Benacerraf 1981; Snell 1981).

Parallel to the immunogenetic studies in the mouse, advances in clinical transplantation led to the discovery of a system analogous to H-2 in man, termed the HLA (human leukocyte antigens) region. Antibodies against leukocytes were described in the sera of polytransfused individuals, multiparous women, and volunteers immunized with leukocytes; these alloantisera were shown to have many different antigenic specificities (Bach and van Rood 1976a,b,c). Studies of natural recombinants in families, and also population genetics, mapped the leukocyte antigens to closely linked loci (HLA-A, HLA-B, and much later, HLA-C) and indicated that they were related to the alloantigens defined in tissue transplantation (for review, see Dausset 1981).

The human equivalent of the mouse I region (HLA-D) was defined as the region controlling the response in the mixed lymphocyte reaction (MLR). This test involved culturing allogeneic lymphocytes and measuring the proliferative response. Evidence from HLA-A and HLA-B-identical siblings that were recombinant for the MLR suggested that the locus determining MLR stimulation was linked to, but distinct from, the established HLA loci (Yunis and Amos 1971). Further analysis of recombinants in families mapped HLA-D centromeric of HLA-A and HLA-B (Reinsmoen et al. 1977). Using the MLR, a series of alleles apparently at a single locus were defined (Bach and van Rood 1976a,b,c). Serological studies identified alloantisera that recognized determinants that cosegregated with the MLR-determined HLA-D specificities; these determinants became known as the HLA-DR (HLA-D related) antigens. The D region has been further subdivided (into HLA-DR, -DQ, and -DP subregions) on the basis of the genetic analysis of serological phenomena and cellular typing (for review, see Travers and McDevitt 1987; Duquesnoy and Trucco 1988). The DR and DQ subregions have not been separated by analysis of recombinants, but the DP subregion was mapped centromeric of DR and DQ (Shaw et al. 1981). In addition, these analyses indicated that the genetic distance between HLA-A and HLA-DP was 3–4 cM.

Loci for the complement components C2, Factor B, and C4 have also been linked to the human MHC. Allen (1974) demonstrated that the electrophoretic polymorphism of Factor B segregated with the HLA alleles. Deficiencies of C2 and C4 in plasma linked these proteins to the MHC (Fu et al. 1974; Hauptmann et al. 1974), and this was confirmed by studies of electrophoretic variants (for review, see Alper 1981). The electrophoretic patterns observed for C4 also indicated that there were two C4 loci (O'Neill et al. 1978), now referred to as C4A and C4B. Another protein whose gene was linked to the human MHC is the enzyme steroid 21-hydroxylase (CYP21), deficiency of which causes congenital adrenal hyperplasia (Dupont et al. 1977). Additional studies of families with intra-HLA recombinant haplotypes confirmed this linkage and suggested that the CYP21 deficiency gene was located between HLA-B and HLA-DR (Dupont et al. 1980).

One striking feature of the products of the class I and class II loci of the MHC is their extreme polymorphism. For example, at least 61 alleles have been described for HLA-B, 41 for HLA-A, 18 for HLA-C, and 72 for HLA-DR (Bodmer et al. 1992). The C4 loci in the class III region also display extensive polymorphism; up to 35 alleles have been recognized (Campbell et al. 1990; Mauff et al. 1990). At most of the polymorphic HLA loci, the heterozygosity is about 90% and most of the alleles occur at a frequency of less than 0.15. Family studies have shown that recombination in the HLA region is rare (<1%), and thus a complete set of alleles are usually inherited as a unit. Such combinations of HLA alleles are referred to as a haplotype (Ceppellini et al. 1967). In addition, population studies have shown that certain combinations of HLA alleles occur together more often than would be expected for a random mating population at Hardy-Weinberg equilibrium. For instance, in American Caucasians, HLA-A1 occurs at a frequency of 0.257, HLA-B8 occurs at a frequency of 0.171, and HLA-DR3 occurs at a frequency of 0.222. The observed frequency of the A1, B8, DR3 haplotype in this population is 0.071, which is significantly different from the expected value of 0.0098 (Tiwari and Terasaki 1985; Bodmer et al. 1987). This phenomenon is known as linkage disequilibrium. The degree of linkage disequilibrium between alleles also depends on the population studied (Bodmer et al. 1987). In addition, linkage disequilibrium has also been shown to include combinations of the complement alleles, leading to the idea of the extended haplotype (Adweh et al. 1983). Linkage disequilibrium may be the consequence of natural selection for or against a specific gene combination, or it may be due to the fact that the population has not yet reached equilibrium. For very closely linked genes, the rate of approach to equilibrium is slow. The stable relationships between alleles has been taken to suggest that these combinations represent preserved ancient or ancestral haplotypes (Dawkins et al. 1983).

PHYSICAL MAPPING OF THE MHC

Characterization of the MHC at the molecular level is of major interest for several reasons. First, it represents ~1/750 of the human genome, and many of its products are well understood. Second, there are a large number of diseases of both an autoimmune and a nonimmune etiology where a genetic predisposition can be mapped to the MHC (Tiwari and Terasaki 1985; Batchelor and McMichael 1987; Hansen and Nelson 1990; Sinha et al. 1990; Todd 1990; Kostyu 1991; Morgan and Walport 1991). In some cases, disease susceptibility can be directly associated with the products of the class I or class II alleles, or with the complement or TNF loci in the class III region. In other cases, however, the correlation may be the result of linkage disequilibrium between the HLA marker and other as-yet-undefined loci within the MHC.

By 1985 cDNA and genomic clones were available for the class I HLA-A, -B, and -C gene products (for review, see Strachan 1987), the class II DR, DQ, and DP α and β chains (for review, see Trowsdale 1987), the complement proteins C2, C4, and Factor B, and the CYP21 enzyme (for review, see Campbell et al. 1986), and the cytokines TNFα and β (for review, see Cerami and Beutler 1988). In addition, overlapping cosmids spanning ~120 kb of DNA had resulted in the alignment of the C2, Factor B, C4A, C4B, and CYP21 genes in the class III region (Carroll et al. 1984, 1985a). In the class II region, the DP, DQ, and DR subregions each contain at least one functional A and B gene which encode the α and β chains, respectively, of the class II molecule, together with a variable number of pseudogenes (see Fig. 1). Overlapping cosmids spanning ~100 kb established the alignment of three DRB genes on the DR3 haplotype (Rollini et al. 1985), and cosmids spanning ~130 kb established the linkage of one DRB gene with the DRA gene on the DR4 haplotype (Spies et al. 1985). In addition, cosmids spanning ~100 kb in the DP subregion and ~83 kb in the DQ subregion established the alignment of two DP α- and β-chain genes (Servenius et al. 1984; Trowsdale et al. 1984) and two DQ α- and β-chain genes (Okada et al. 1985), respectively.

The availability of such a large number of well-characterized loci spread over about 3–4 cM (which is very approximately equivalent to 3000–4000 kb) meant that the MHC was a suitable system for the application of large DNA restriction fragment mapping using PFGE (Carle and Olson 1984; Schwartz and Cantor 1984; Southern et al. 1987). The first physical map of any region of the MHC was established for the class II region in 1986 using DNA from a cell line of the DR4 haplotype

(Hardy et al. 1986). This established the relative order of the subregions and a linkage map of all the known class II genes at that time. The DP subregion, the most centromeric class II subregion, was separated by ~300 kb from the DQ loci, which in turn were separated from the DRB genes by ~120 kb. There was a gap of ~100 kb between the DRB and DRA loci. A single α-chain gene, DNA, was placed telomeric of the DP subregion within ~75 kb, and a single β-chain gene, DOB, was placed ~90 kb telomeric of the DQ subregion. The PFGE analysis indicated that on the DR4 haplotype, the DRA gene was separated from the DPB2 gene by ~900 kb of DNA. This analysis also indicated that the DR-DQ distance was similar to the DQ-DP distance. This is of particular interest, because no recombination between the DQ and DR genes has yet been reported, and alleles of these genes are in very strong linkage disequilibrium. In contrast, a high recombination frequency of ~1-3% occurs between the DP genes and the DQ-DR genes. Furthermore, there is little linkage disequilibrium between DP and DR alleles. Because the distance between DR and DQ is similar to the distance separating DQ and DP, this suggests that a recombination "hot spot" exists between the DP and DQ-DR subregions.

This map was subsequently extended in 1987 by Dunham et al. (1987), Carroll et al. (1987), and Lawrance et al. (1987) to include the class III and class I regions and to produce the first physical maps of the whole MHC. In the case of Dunham et al. (1987) and Carroll et al. (1987), genomic DNA from the HLA homozygous consanguineous cell line ICE5 (HLA type: A2 Cw7 B7 C2C BfS C4A3 BQO DR2) was used to minimize mapping problems caused by possible haplotype-specific RFLPs. The total size of the MHC hybridizing DNA fragments detected suggested that the MHC spanned ~4000 kb of DNA, which represents ~1/750 of the human genome; this is in good agreement with the estimate for the size of the complex of 3-4 cM from recombination studies in families. The class II genes at the centromeric end are contained within 850 kb, whereas the class I genes at the telomeric end are contained within ~2000 kb of DNA. The most telomeric class II gene, DRA, and the most centromeric class I gene, HLA-B, are separated by ~1100 kb of DNA. These mapping experiments also established the orientation and position of the complement and TNF gene clusters in the class III region with respect to the class I and class II regions. The DRA to CYP21B distance was estimated to be ~400 kb, the C2 to TNF-A distance was estimated at ~350 kb, and the TNF-B to HLA-B distance was estimated to be ~220 kb. Further refinements of the original maps have been carried out by a number of different groups, including Dunham et al. (1989a, 1990), Ragoussis et al. (1988, 1989), Tokunaga et al. (1988), Lord et al. (1990), Kendall et al. (1991), Shukla et al. (1991), and Kahloun et al. (1992), allowing better estimates of the distances between the various loci. In particular, more detailed physical maps of the class I region are available allowing the location of various class I loci to be established. A physical map of the MHC constructed from the data derived from the PFGE studies is given in Figure 1. In general, the physical maps have agreed with each other, the most common problem being variation in fragment size estimation. Differences in observed fragment sizes can be explained by variation of the PFGE systems used, by haplotype-specific restriction fragment length polymorphisms (RFLPs), or by insertions or deletions of DNA. There could also be methylation differences between cell lines leading to some sites not being recognized, since the enzymes used in PFGE analysis are sensitive to cytosine methylation at the CpG dinucleotides in their recognition sequences. For instance, although a *Nru*I site centromeric to the TNF genes was observed to be cleaved in two cases (Carroll et al. 1987; Dunham et al. 1987), it was not seen in another case (Ragoussis et al. 1989).

Tokunaga et al. (1988) have carried out comparative analysis of six different haplotypes by PFGE and have observed a number of restriction fragment length differences. However, multiple examples of each haplotype showed specific genomic characteristics, including deletions, duplications, or insertions, supporting the hypothesis that the haplotypes studied represent conserved ancestral haplotypes. PFGE has also been used to estimate gene copy number carried by different MHC haplotypes (Dunham et al. 1989b; Zhang et al. 1990), to ascertain the extent of restriction fragment length variation in genomic DNA prepared from blood samples of an unselected population (Lawrance and Smith 1990), and to carry out comparative mapping of the MHC in different racial groups (Tokunaga et al. 1989).

PFGE analysis of the class II region Comparative analysis of restriction site mapping data in a number of different class II haplotypes by PFGE has suggested that differences exist in the DNA organization between haplotypes (Tokunaga et al. 1988; Dunham et al. 1989a; Inoko et al. 1989; Kendall et al. 1991). The DNA organization in the DR3, 5, and 6 haplotypes is the same, and the distance between the DRA and DPB2 genes is ~750 kb (Fig. 2). However, in cell lines possessing the DR2 haplotype, evidence has been obtained for the presence of an extra 20–30 kb of DNA in the DRB region when compared to the DR3, 5, and 6 haplotypes (Dunham et al. 1989a). Similarly, the DR4 haplotype appears to have a large additional segment of DNA (~120 kb), irrespective of subtype, compared to the DR3, 5, and 6 haplotypes in the region containing the DRB genes (Fig. 2). In addition, a number of groups have studied the DR7 and DR9 haplotypes, which are related to the DR4 haplotype by virtue of sharing a particular supertypic specificity (DRw53), and have obtained results consistent with ~120 kb more DNA present in the region containing the DRB and DQA genes (see Fig. 6) (Dunham et al. 1989a; Inoko et al. 1989; Tokunaga et al. 1989; Kendall et al. 1991). Therefore,

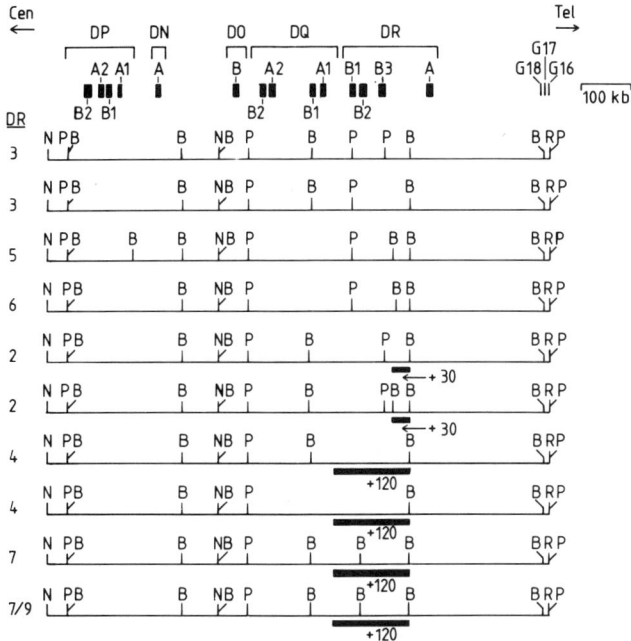

Figure 2 Comparative restriction maps of a number of class II haplotypes. The maps have been constructed from data derived from PFGE studies (Dunham et al. 1989a; Kendall et al. 1991). Enzyme sites shown are for B, *Bss*HII; N, *Not*I; P, *Pvu*I; R, *Nru*I. The class II genes are shown at the top of the figure as black boxes. Those genes indicated by bars are the G18, G17, and G16 genes in the class III region. Black bars indicate the position where the DNA content varies between haplotypes. The extent of the variation in kb is also given.

there appears to be consistency in the observations of the amount of DNA present in cell lines possessing the same haplotype. However, a number of variations in the positions of cleavable restriction sites between even the related cell lines might be due to methylation or DNA sequence differences (Fig. 2).

It has been estimated that there are at least three DRB genes on each of the haplotypes DR2 to DR6, whereas the DR4, 7, and 9 haplotypes possess four DRB genes (see Fig. 7) (see Trowsdale et al. 1991). It is therefore possible that the size variation between the haplotypes is due to differences in the DRB gene organization, and/or because of the differences in the numbers of duplicated DRB genes. This would seem particularly likely in the DR2 case, where the difference of 20–30 kb is localized to the region containing the DRB genes. However, another possible cause for the differences is that there are variations in the amount of DNA present in the DR subregion between haplotypes.

This would seem likely in the case of the DR4, 7, and 9 haplotypes because an alteration of ~120 kb cannot solely be caused by the presence of an extra DRB gene, since the organization of the DRB genes in the cloned cosmid DNA (Spies et al. 1985) is not consistent with such an insertion. These differences may be of significance for the maintenance of linkage disequilibrium within the DR and DQ subregions and for the lack of recombination events between DR and DQ.

PFGE analysis of the class III region Phenotypic genetics have established that null alleles at the C4 loci are relatively common with gene frequencies for C4A null alleles (QO) of 5–15% and for C4BQO alleles of 10–20% (Schendel et al. 1984; Tokunaga et al. 1985; Partanen and Koskimies 1986). For C4AQO alleles in 50% of cases, absence of the C4A gene product is due to gene deletion usually together with the flanking CYP21 gene (Carroll et al. 1985b). In the case of the C4BQO phenotype, gene deletion accounts for ~10% of cases. In addition, duplication of the C4 genes, including the flanking CYP21 genes, has also been observed, and haplotypes with three and even four C4 genes have been reported (Raum et al. 1984; Rittner et al. 1984; Uring-Lambert et al. 1984; Collier et al. 1989). Variation in size of the C4 gene at locus II (C4B) has also been observed, and the gene can be either 22 kb or 16 kb in length (Prentice et al. 1986; Yu et al. 1986). The basis of the size difference is in the length of the intron that separates exons 9 and 10. In long genes this intron is ~6.5 kb in length (Yu 1991), whereas in short genes the intron is only ~0.35 kb in length (M.J. Anderson and R.D. Campbell, unpubl.). Some of the gene organizations that have been established at the C4 loci are summarized in Figure 3. The different gene organizations can be observed directly using PFGE and the enzymes *Bss*HII (Dunham et al. 1989b; Collier et al. 1989) and *Ksp*I (E. Kendall and R.D. Campbell, unpubl.), since the size of the restriction fragment generated by each enzyme correlates directly with the number and length of the C4 genes present within the fragment (Fig. 3).

Comparative analysis of a number of different haplotypes has revealed that there are no gross differences in the DNA organization between haplotypes elsewhere in the class III region apart from the known differences in C4 and CYP21 gene number (Dunham et al. 1990).

PFGE analysis of the class I region The class I region spans ~2000 kb of DNA at the telomeric end of the MHC. This region contains the HLA-A, -B, and -C genes that encode the highly polymorphic α chains of the class I cell-surface molecule (Strachan 1987). Recently, a number of related genes have been identified by screening cosmid and cDNA libraries with class I probes at low stringency. Three of these, designated HLA-E, HLA-F, and HLA-G, encode intact proteins that are known to be ex-

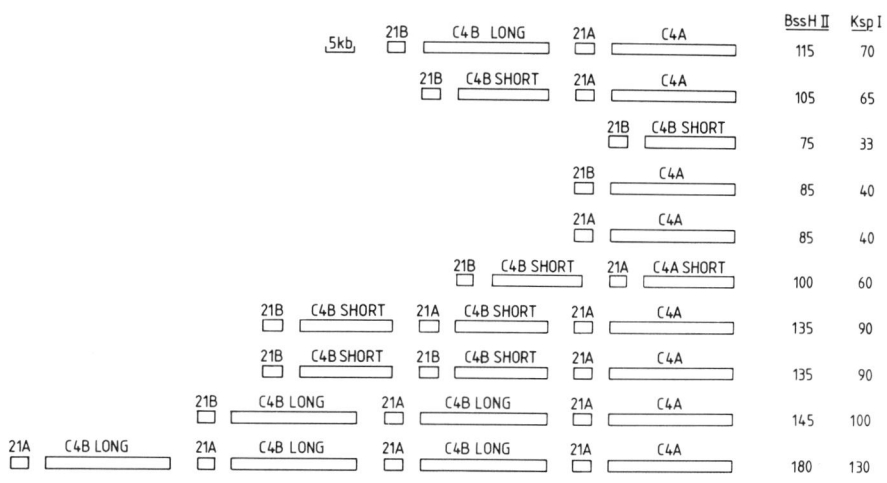

Figure 3 Some of the gene organizations of the C4 and CYP21 loci in the class III region. The sizes of the *Bss*HII and *Ksp*I restriction fragments observed in peripheral blood DNA by PFGE are shown at the right of the figure and are diagnostic for C4/CYP21 gene number and size of the C4 genes (Dunham et al. 1989b; E. Kendall and R.D. Campbell, unpubl.).

pressed (Geraghty et al. 1987, 1989; Koller et al. 1988). However, there is no evidence for expression of the HLA-H gene at the protein level. The HLA-J gene (previously designated the cda 12 gene), which encodes a product quite distinct from the class I α chains, has been identified ~50 kb from HLA-A (Ragoussis et al. 1989; Messer et al. 1992).

The arrangement of the class I loci has been studied by analysis of recombination within informative pedigrees (Orr and DeMars 1983), and of irradiation-induced HLA-loss mutants (Koller et al. 1989), as well as by PFGE. The PFGE analysis has revealed that the relative order of the genes is roughly consistent with that obtained by recombination analysis. The HLA-B and -C genes are ~130 kb apart (Pontarotti et al. 1988). However, because of the large distance separating the HLA-C and -A loci and the lack of informative probes, it proved difficult to link these loci by PFGE. Recently, through the isolation of YAC clones from this region, it has been possible to isolate suitable probes that have aided in the construction of the physical map (Figs. 1 and 7) (Chimini et al. 1990; Bronson et al. 1991; Geraghty et al. 1992c; Kahloun et al. 1992). This has revealed that the HLA-A gene lies ~1400 kb telomeric to HLA-C. Of the nonclassical class I genes, the HLA-E gene lies ~700 kb from HLA-C, whereas the HLA-G, HLA-F, and HLA-H genes all map telomeric of HLA-A (Shukla et al. 1991; Schmidt and Orr 1991). By recombination analysis, HLA-G and HLA-F were thought to lie ~8 cM telomeric to HLA-A, yet physical mapping indicates that the two genes are within 250 kb of HLA-A. The large discrepancy between the genetic distance and the physical distance between these genes suggests that there is a recombination hot spot telomeric of HLA-A. PFGE analysis using HLA-A, -B, and -C locus-specific probes has shown that the DNA content of the class I region in different haplotypes is relatively invariant (Chimini et al. 1988). This study reported that, despite the high nucleotide polymorphism in this region, no RFLPs were observed between the different haplotypes using the rare-cutting endonuclease *Sfi*I, and only two haplotypes showed variant bands with *Mlu*I. However, variation in the size of *Not*I fragments detected using an HLA-A probe have been reported (Lawrance and Smith 1990; Lord et al. 1990). Further analysis has indicated that individuals expressing the HLA-A24 and -A3 alleles possess haplotype differences of up to 90 kb between HLA-E and HLA-A compared to other HLA-A alleles (Fig. 4) (Kahloun et al. 1992). In addition, the HLA-A24 and HLA-A23 haplotypes possess a ~50-kb deletion that falls between the HLA-A and HLA-G genes and specifically includes the HLA-H gene compared to other HLA-A haplotypes (Venditti and Chorney 1992).

MOLECULAR MAP

Overlapping cosmid and YAC clones are now available which cover the entire MHC. The cosmid and YAC clones were isolated either by using the available class I, class II, and class III region probes, or by standard techniques of chromosome walking. These have allowed highly detailed molecular maps of the MHC to be constructed and have aided greatly in the search for novel genes within the MHC. These genes have been lo-

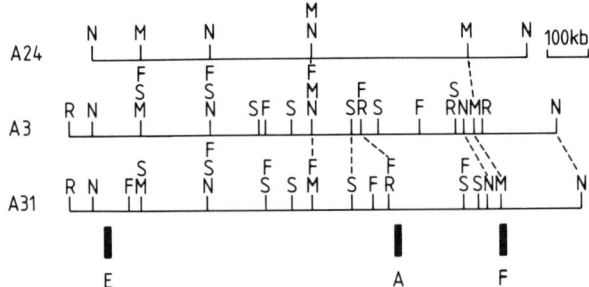

Figure 4 Comparative restriction maps of three class I haplotypes. The restriction maps have been constructed from data derived from PFGE studies. Enzyme sites shown are for F, *Sfi*I; M, *Mlu*I; N, *Not*I; R, *Nru*I; S, *Sal*I. The HLA-E, -A, and -F genes are illustrated by black boxes. Equivalent restriction sites between the haplotypes around the HLA-A and -F genes are indicated by dotted lines. (Adapted from Kahloun et al. 1992.)

cated using a variety of different techniques, including HTF-island mapping, zoo blot analysis, the direct screening of cDNA libraries using cosmid genomic inserts, and most recently, cDNA selection on immobilized YAC DNA and recovery of the selected cDNAs by polymerase chain reaction (PCR). The methods that have been used for the identification of novel genes are described in detail with reference to the class III region.

Novel genes in the class III region Sargent et al. (1989a,b), Kendall et al. (1990), and Spies et al. (1989a,b) have isolated a series of overlapping cosmid clones covering an ~890-kb region extending from the HLA-B locus to a point ~200 kb telomeric of the HLA-DRA locus (Fig. 5). The remaining ~200 kb has been cloned in YAC clones (Ragoussis et al. 1991).

One way to define the location of a gene in the cloned DNA is to look for evolutionary conservation of sequence. Since transcribed sequences are more highly conserved in evolution than noncoding sequences, genomic DNA fragments that cross-hybridize with the DNA of other species in genomic Southern blot analysis can be taken to indicate the presence of potential coding sequences (Monaco et al. 1986). This is usually referred to as zoo blot analysis. Using this strategy, Levi-Strauss et al. (1988) were able to locate a novel gene called RD between the Factor B and C4 genes. In their characterization of the class III region, Sargent et al. (1989a,b) used a number of probes isolated from cosmid genomic clones spanning the gap between the C4 and TNF genes in zoo blot analysis. All of the probes used were found to cross-hybridize with DNA from the other mammalian species and one (probe H) also cross-hybridized with shark DNA. In all cases where cross-hybridization was observed, the probe was subsequently shown to include, or be part of, a novel gene, reinforcing the notion that this type of analysis is an excellent indicator of the presence of a potential coding sequence. For example, detailed characterization of the region around probe H has led to the identification of three genes encoding members of the major heat shock protein HSP70 family (Sargent et al. 1989b; Miner and Campbell 1990, 1992).

One major consequence of the PFGE studies described above has been the identification of a large number of sites in the MHC, and in particular in the class III region, for rare-cutting restriction endonucleases with one or two CpG dinucleotides in their recognition sequence (e.g., *Bss*HII, *Sac*II [or its isochizomer *Ksp*I], *Eag*I, and *Not*I). The sites for many of these enzymes are not randomly distributed but are clustered at CpG islands commonly referred to as HTF (*Hpa*II tiny fragment) islands (Bird 1987). These are short stretches (0.5–2 kb) of DNA that (unlike the rest of the human genome) are not depleted in unmethylated CpG dinucleotides and are often associated with the 5′ ends of housekeeping and some tissue-specific genes (Bird 1987: Gardiner-

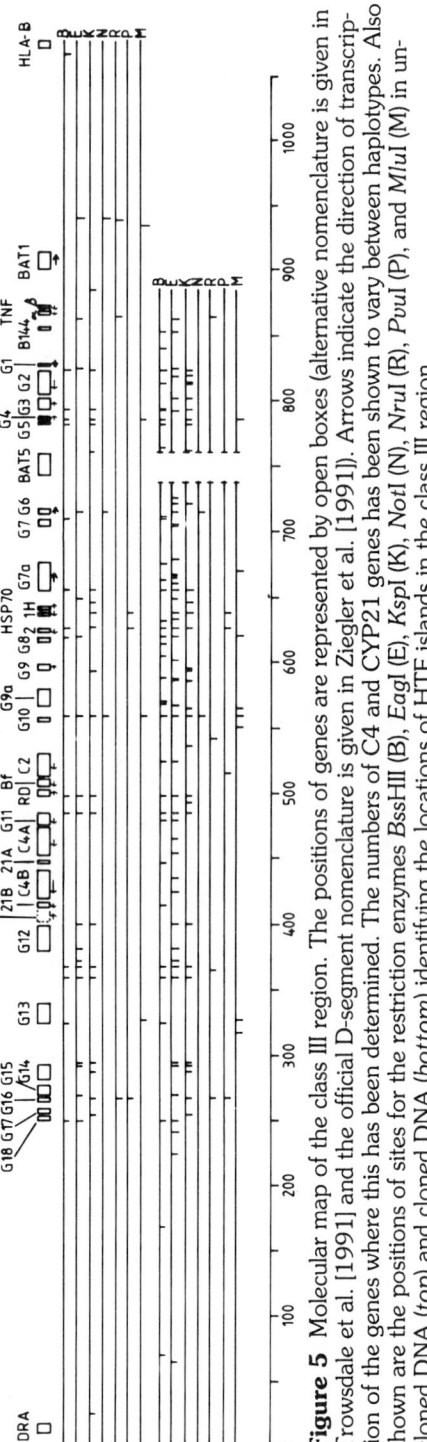

Figure 5 Molecular map of the class III region. The positions of genes are represented by open boxes (alternative nomenclature is given in Trowsdale et al. [1991] and the official D-segment nomenclature is given in Ziegler et al. [1991]). Arrows indicate the direction of transcription of the genes where this has been determined. The numbers of C4 and CYP21 genes has been shown to vary between haplotypes. Also shown are the positions of sites for the restriction enzymes *Bss*HII (B), *Eag*I (E), *Ksp*I (K), *Not*I (N), *Nru*I (R), *Pvu*I (P), and *Mlu*I (M) in uncloned DNA (*top*) and cloned DNA (*bottom*) identifying the locations of HTF islands in the class III region.

Garden and Frommer 1987; Larsen et al. 1992). Restriction mapping of the cosmid cloned (i.e., unmethylated) DNA identified a large number of sites for these enzymes, and the methylation status in uncloned genomic DNA was established by PFGE analysis (Sargent et al. 1989a; Kendall et al. 1990). This analysis identified 33 clusters of sites, or single sites, for these enzymes containing unmethylated CpG dinucleotides (Fig. 5).

In the region separating the C4 and TNF genes, Sargent et al. (1989a) successfully used genomic probes adjacent to or including the HTF islands to detect mRNAs on Northern blots of total RNA from a panel of cell lines. Subsequently, cDNA clones corresponding to these novel genes, designated G1–G11 (Fig. 5), have been isolated. Another approach used to locate genes is to isolate cDNA clones by hybridizing whole cosmid genomic inserts onto a cDNA library using a preannealing procedure to prevent cross-hybridization by repetitive DNA elements. This allows the possibility of identifying the maximum number of coding regions, including non-HTF-island-associated genes, with the minimum number of probings of the cDNA library. Once isolated, the cDNA inserts can be mapped back onto the relevant cosmid insert to define the location of the corresponding gene. This strategy has been successfully used by Kendall et al. (1990) to locate seven novel genes, G12–G18, in a 160-kb segment of DNA extending centromeric of the C4 genes, at least five of which are associated with HTF islands (Fig. 5). In their analysis of the class III region, Spies et al. (1989a,b, 1990) have made extensive use of whole cosmid genomic inserts to directly screen a T-cell-derived cDNA library, and this has led to the identification of nine novel genes, designated BAT1–BAT9, between C2 and HLA-B, and six novel genes, designated X1–X6, between CYP21 and HLA-DRA. Apart from BAT1 and BAT5 (Fig. 5), the remainder of the novel genes isolated by Spies et al. (1989a,b, 1990) correspond to genes isolated by Sargent et al. (1989a,b) and Kendall et al. (1990).

In addition, Morel et al. (1989) have located a gene immediately adjacent to the CYP21B gene (Fig. 5). In this case, the gene is encoded on the opposite strand to that encoding CYP21 and overlaps with the 3′ end of the CYP21 gene. This novel gene, labeled XB or OSG, is also part of the unit of duplication at the C4 loci such that part of the sequence of the OSG is duplicated and is contained between the CYP21A and C4B genes (Gitelman et al. 1992). This gene has been labeled XA and is also expressed at the mRNA level.

Thus far, 38 genes have now been located in a 680-kb segment of DNA in the class III region. Half of these (18) are associated with HTF islands and appear to be ubiquitously expressed. However, intermingled with these genes are others which are expressed in a more tissue-specific manner. For example, the principal site of synthesis of the complement proteins C4, C2, and Factor B is the liver (Colten and Dowton 1986), whereas the CYP21B and XA genes are only expressed in the adrenal gland (Morel et al. 1989). In addition, one of the novel genes, G1, defined by Sargent et al. (1989a) appears to have a more restricted expression, the G1 mRNA being observed in cell lines representing monocytes (U937) and T lymphocytes (Molt 4). Thus the class III region, and particularly the ~240-kb gap between the DRA and G18 genes and the ~170 kb between the BAT1 and HLA-B genes, may contain several more genes, many of which could be expressed in a tissue-specific manner. This is emphasized by the recent finding of Olavesen et al. (1993a) who have identified a novel gene, G7b, of ~10 kb in length in the 13-kb gap separating the HSP70-Hom and G7a genes.

One consequence of having so many genes in such a small segment of DNA is that the genes in some parts of the class III region are very closely spaced. For example, the 5'end of the Factor B gene lies only 421 bp from the polyadenylation site of the C2 gene (Wu et al. 1987); the 3' end of the Factor B gene is within 500 bp of the 3'end of the RD gene (Campbell and Porter 1983; Speiser and White 1989); and the OSG and CYP21 genes actually overlap by 481 bp at their 3'ends (Morel et al. 1989). Thus, it is possible to imagine that transcription of one gene could interfere with the transcription of a closely linked gene. Recently, it has been shown that the 5'-flanking region of the Factor B gene contains a sequence motif, ME1a1, which is bound by the MAZ zinc finger protein (Bossone et al. 1992) and which elicits transcriptional termination of the C2 gene (Ashfield et al. 1991). Terminating C2 transcription in this manner helps prevent interference with the Factor B promoter and could be a strategy utilized by other closely spaced genes transcribed by RNA polymerase II.

Novel genes in the class II region The largest contiguous set of overlapping cosmids in the class II region corresponds to a 280-kb segment linking the DOB gene to the DQ subregion (Blanck and Strominger 1988). Other smaller portions of the class II region around the DRB genes (Rollini et al. 1985; Spies et al. 1985), around the DNA gene (Blanck and Strominger 1990), and in the DP subregion (Servenius et al. 1984; Trowsdale et al. 1984) have also been isolated in overlapping cosmids, and a cosmid walk extending ~100 kb centromeric of the DPB2 locus has led to the identification of the COL11A2 gene 45 kb from DPB2 (Hanson et al. 1989). These have now been linked in a series of overlapping YAC clones (Ragoussis et al. 1991; Kozono et al. 1991), completely encompassing the class II region in a 1200-kb contig and extending into the class III region (Ragoussis et al. 1991).

Detailed mapping of the available cosmids identified a large number of sites for rare-cutting enzymes in the cloned DNA, and those which also cleaved in uncloned DNA were established by PFGE (Hanson and Trowsdale 1991; Hanson et al. 1991). This identified 14 clusters of sites or single sites that potentially correspond to HTF islands (Fig. 6). Five of

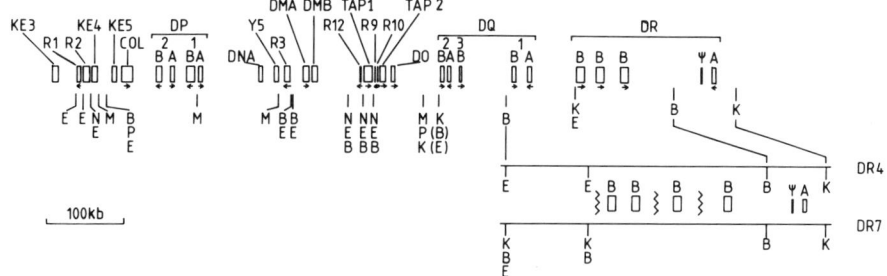

Figure 6 Molecular map of the class II region. The positions of genes are represented by open boxes (alternative nomenclature is given in Trowsdale et al. [1991] and the official D-segment nomenclature is given in Ziegler et al. [1991]). Arrows indicate the direction of transcription of the genes where this has been determined. The number of DRB genes has been shown to vary between haplotypes. Also shown are the locations of sites for the restriction enzymes BssHII (B), EagI (E), KspI (K), MluI (M), NotI (N), and PvuI (P) found to cleave in uncloned genomic DNA. The location of the differences in DNA content between the DR4 and DR7 (and DR9) haplotypes and the DR3 haplotype is shown. (Adapted from Kendall 1992.)

these lie at the centromeric boundary of the MHC just centromeric of the COL11A2 gene, and 3 are grouped in a region ~30–60 kb telomeric to the DNA gene. Three potential HTF islands lie in a 35-kb region ~20 kb from the DOB gene, and two more are located between the DOB and DQB2 genes. Subsequent use of genomic fragments or whole cosmid inserts by Trowsdale and colleagues to screen cDNA libraries has resulted in the isolation of a large number of cDNAs. These have been mapped back onto the cosmids, thus locating 12 novel genes designated RING (really interesting new genes) 1–12 (Fig. 6) (Trowsdale et al. 1990; Glynne et al. 1991; Hanson et al. 1991; Kelly et al. 1991a,b). Spies et al. (1990) have also used cosmid inserts derived from the region between the DNA and DOB2 genes to screen a T-cell-derived cDNA library, and they have identified five novel genes, Y1-Y5. Of these, Y5, which lies just telomeric of DNA, was not detected by Trowsdale and colleagues (Fig. 6). In addition, probes from non-class-II genes in the mouse H-2K region have been used to map human equivalents of the murine KE3-KE5 genes (Abe et al. 1988) to the region centromeric of the DP subregion (Hanson and Trowsdale 1991). KE4 was found to correspond to RING5, whereas KE3 and KE5 represent previously unidentified genes in this region (Fig. 6).

As for the class III region, the number of genes so far located in the class II region will represent the minimum number of genes in this region. At least two potential HTF islands, one between the DOB and DQB2 genes and one between the RING12 and DMB genes (Fig. 6), have yet to have genes assigned to them. In addition, there are gaps of sufficient size between the known loci that could potentially contain many more novel genes.

Novel genes in the class I region A number of studies have shown that there are at least 18 distinct class-I-like gene sequences, each lying within a separate *Hin*dIII fragment (Koller et al.1989; Lawlor et al. 1990). Six of these encode the classical transplantation antigens HLA-A, -B, and -C and the less polymorphic HLA-E, -F, and -G. Of the 12 remaining sequences, 4 are full-length pseudogenes and 8 are abbreviated pseudogenes, including 3 containing a single intron-exon fragment (Geraghty et al. 1992a,b,c). Efforts to clone the class I region in cosmid vectors have not resulted in the isolation of large contiguous regions because of the comparatively large size of the class I region. Recently, however, YAC cloning has been successfully used to isolate a 290-kb portion of the class I region containing the HLA-B and HLA-C loci and an HLA-B-linked pseudogene (Bronson et al. 1991), a 355-kb portion containing HLA-E and a class I pseudogene designated HLA-X (Chimini et al. 1990), and a 1200-kb portion linking HLA-E and HLA-F at the telomeric end of the MHC that includes 14 of the 18 characterized class I sequences (Geraghty et al. 1992c). Restriction enzyme mapping and the use of locus-specific probes have allowed all of the class I genes and sequences to be ordered and positioned within the region (Fig. 7). In addition, the transcriptional orientation of the four class I genes has been determined. Studies using hybridization with YAC insert end probes have identified a region between the HLA-A and HLA-G genes that varies in size in certain HLA haplotypes (Geraghty et al. 1992c; Kalhoun et al. 1992; Venditti and Chorney 1992). It is of considerable interest that cosmid and YAC clones that contain DNA from this region are unstable and rearrange or shorten upon propagation. Also of interest and perhaps relevant to this instability is the contrast between the relatively high recombination distance between the HLA-A and HLA-F loci (Koller et al. 1989) and a physical distance of only ~250 kb. With the region cloned and with an extensive array of probes in hand, it should be a simple matter to narrow down the sites of recombination and ultimately identify a potential recombination sequence in the human genome.

Within the YAC clones there are 13 positions where two or more occurrences of the CpG-containing sites cleaved by restriction enzymes

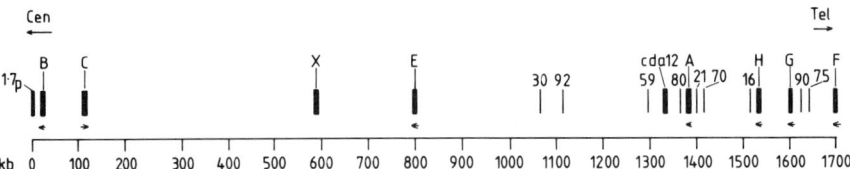

Figure 7 Molecular map of the class I region. Those genes designated by a letter (e.g., HLA-A, HLA-B, etc.) are indicated by black boxes. Bars indicate the location of 10 class-I-like pseudogene or gene-fragment sequences, which are illustrated by their numerical designation (Koller et al. 1989).

*Bss*HII, *Mlu*I, *Not*I, and *Sal*I appear within 5 kb of one another and not in association with a class I locus (Chimini et al. 1990; Geraghty et al. 1992c). A number of these have been shown to be unmethylated in uncloned genomic DNA (Kalhoun et al. 1992). Thus, these clustered sites probably correspond to HTF islands, and since HTF islands are almost always associated with transcribed sequences, it is likely that they define the location of novel genes. Recently, Parimoo et al. (1991) and Lovett et al. (1991) have developed a novel technique for identifying the location of genes in cloned YAC DNA. This method involves the hybridization of cDNA fragments, generated from a total cDNA library by PCR, to the immobilized YAC DNA and recovery of the selected cDNAs by further rounds of PCR. By use of this approach, Parimoo et al. (1991) have been able to identify several non-class-I cDNAs from a 320-kb YAC that includes the HLA-A locus. Thus, it is clear that novel, non-class-I genes are located within the class I region. If the gene density in the class I region is comparable to the gene density in the class III region, then upward of 50 novel genes could be located in the 2000 kb of DNA that constitutes the class I region.

WHAT DO THE GENES ENCODE?

The class I and class II genes, and the complement component and TNF genes in the class III region, all encode proteins that perform different immune-related functions. It is likely, therefore, that some of the novel genes in the MHC also encode proteins that perform functions that are important for the immune response. Characterization of the novel genes by cDNA sequence analysis has so far revealed three types of novel genes:

1. *New class II genes.* The RING6 and RING7 genes in the class II region (Fig. 6) (now designated DMA and DMB, respectively) have recently been shown to encode molecules that are related to the classical HLA antigens (Kelly et al. 1991a). RING6 shows similarity to the class I α-chain genes and β2-microglobulin. RING7 is equally related to class I and class II genes, but does contain sequence motifs characteristic of class II molecules.
2. *Genes with a possible immune function such as peptide generation and transport.* The predicted protein products of the closely linked RING4, RING10, RING11, and RING12 genes in the class II region (Fig. 6) are particularly interesting. The RING4 (Y3 or PSF1) and RING11 (Y1 or PSF2) genes encode highly related protein products (748 and 703 amino acids, respectively) that are members of the ATP-binding cassette superfamily of transporters (Parham 1990; Spies et al. 1990; Trowsdale et

al. 1990; Bahram et al. 1991; Spies and DeMars 1991; DeMars and Spies 1992). An antiserum raised specifically against RING4 has been used to demonstrate that the RING4 and RING11 proteins assemble to form a complex (Kelly et al. 1992; Spies et al. 1992). In addition, experiments with mutant cell lines have shown that defects in either of these proteins result in the lack of cell-surface expression of class I molecules and thus prevent the presentation of intracellular antigens (Kelly et al. 1992; Spies et al. 1992). This supports the hypothesis that RING4 and RING11 are the components of a transmembrane protein complex that plays a crucial role in the import of antigenic peptides from the cytosol to the lumen of the endoplasmic reticulum. The RING10 (Y2) and RING12 genes (now designated LMP2 and LMP7, respectively) encode proteins of 272 and 219 amino acids, respectively, that show sequence similarity with proteasome components (Glynne et al. 1991; Kelly et al. 1991b). Proteasome is a large intracellular multiproteinase complex (Goldberg and Rock 1992), which suggests that the RING10 and RING12 proteins may be important in the degradation of antigens in the cytoplasm. The expression of the putative proteasome and peptide transporter genes is induced by γ-interferon. Thus, it is possible that these genes are coordinately regulated with the genes encoding the class I and class II glycoproteins.

In the class III region, three genes encoding members of the major heat shock protein HSP70 family have been identified (Sargent et al. 1989b; Milner and Campbell 1990, 1992). The members of the HSP70 protein family are known to interact with damaged or denatured proteins and to be involved in the folding, unfolding, and translocation of proteins within the cell (for review, see Langer and Neupert 1991). A member of the rat HSP70 family, prp70, has been shown to bind cytosolic proteins that contain a particular peptide motif and thus mediate their translocation to the lysosomes for degradation (Chiang et al. 1989; Dice 1990). The constitutively expressed HSC70 has been shown to bind a variety of synthetic peptides (Flynn et al. 1989), whereas a mouse HSP70, PBP72/74, has been shown to participate in the binding of antigenic peptides to class II molecules (VanBuskirk et al. 1991). These activities indicate a number of potential roles for the HSP70 proteins in the immune response in relation to antigen processing and presentation (DeNagel and Pierce 1992).

3. *Genes with no known immune function and no obvious relationship with the MHC.* The G7a gene in the class III region encodes a 1265-amino acid protein that shows 48.3% identity

with the valyl-tRNA synthetase of *Saccharomyces cerevisiae* over 1043 amino acids (Hsieh and Campbell 1991). In addition, the G7a protein contains two short consensus sequences characteristic of class I tRNA synthetases and has a predicted molecular mass (~140.5 kD) close to that of other mammalian valyl-tRNA synthetases. Taken together, this is strong evidence that G7a encodes the human valyl-tRNA synthetase. In the class II region, the RING3 gene product is homologous to the *Drosophila* gene, *female sterile homeotic* (*fsh*), as well as to some other proteins that may be involved in cell division (Trowsdale et al. 1991).

What the functions of the other gene products are and whether they are involved in the immune response remains to be established. However, the proteins encoded by some of the other novel MHC region genes do show interesting sequence similarities with known proteins over restricted regions.

G1 encodes a 10-kD protein that contains two potential "EF-hand"-type calcium-binding domains in its amino-terminal half (Olavesen et al. 1993b). The expression of G1 appears to be restricted to T lymphocytes, monocytes, and macrophages, suggesting that it may have a role in the activation of these cell types. The 110-kD protein product of the G9a gene contains six contiguous copies of a 33-amino acid motif known as the cdc 10/SW16, or ankyrin repeat (Milner and Campbell 1993). Repeats of this type may have a role in intracellular protein-protein interactions and have been identified in a large number of proteins with functions that include the regulation of tissue differentiation (e.g., the Notch protein of *Drosophila*) and control of the cell cycle (e.g., SW16 in *S. cerevisiae*) (Blank et al. 1992; Michaely and Bennett 1992). The cysteine-rich carboxy-terminal region of G9a has been shown to bind zinc (Milner 1991). The predicted protein product of G13 shows significant sequence similarity with the leucine zipper family of transcription factors, suggesting that it may have a role in the regulation of transcription (Khanna and Campbell 1993). Partial sequence of the OSG has indicated that this gene encodes a protein which contains a number of repeats corresponding to the consensus for the 90-amino acid fibronectin type-III repeat (Gitelman et al. 1992; Matsumoto et al. 1992a,b). The type-III repeats in OSG are most similar to those in tenascin—an extracellular matrix glycoprotein. Finally, the RING1 gene contains an arrangement of cysteine and histidine residues in a conserved motif, found in a group of proteins including the human V(D)J recombination activating gene RAG-1 (Freemont et al. 1991). This motif is similar to those of zinc finger protein sequences and appears to define a new family of metal-dependent DNA-binding proteins (Reddy et al. 1992).

CONCLUSION

The human MHC now represents one of the most extensively characterized regions of the human genome. The class II and class III regions together span ~2000 kb of DNA and contain at least 70 genes (including some related pseudogenes), both housekeeping and tissue-specific. The gene density is exceptionally high and in some segments, especially in the class III region, is now approaching one gene every few kilobases. There is every expectation that the total number of genes in the class II and class III regions will increase significantly, and there is ample space between the existing genes for at least another 30–50 novel genes. Novel genes have already begun to be located in the class I region which also spans ~2000 kb of DNA, and there is every likelihood that the gene density here is similar to the gene density elsewhere in the MHC. If this proves to be correct, then upward of 200 genes could be located in the MHC.

The human MHC is located in the chromosome band 6p21.3, which is a Giemsa (G)-negative or -light band (Spring et al. 1985; Bickmore and Sumner 1989). A recent survey of gene localizations by in situ hybridizations indicates that the bulk (74%) of genes are located in G-light bands (Bickmore and Sumner 1989), which are also the most GC-rich (Gardiner et al. 1990). The detection of such a large number of genes in the MHC is compatible with the suggestion that genes, both housekeeping and tissue-specific, may predominantly be located in G-light bands. The in situ nick translation of human metaphase chromosomes with *Msp*I and *Hpa*II has shown an association of unmethylated CpG dinucleotides with G-light bands (Adolph and Hameister 1990). In addition, PFGE analyses of different regions of the genome appear to indicate that HTF islands, and therefore HTF-island-associated genes, occur much more frequently in the G-light bands than in G-dark bands. For example, the isolation of *Eag*I/*Eco*RI fragments from a hamster/human hybrid containing the chromosomal region Xq24-qter showed that 30 CpG islands were nonrandomly distributed with respect to the banding pattern (Maestrini et al. 1990). The majority of the CpG island probes mapped into the G-light bands Xq24 and Xq28, whereas none were found in the G-dark band Xq25. On chromosome 11 in the G-light band 11p13, 9 CpG islands, constituting a "CpG island archipelago," have been identified in 850 kb of DNA (Bonetta et al. 1990), whereas a 750-kb segment on chromosome 16 in the G-light band 16p13.3 has been predicted to be very gene-rich (Germino et al. 1993). On chromosome 4 in the G-light band 4p16.3, 15 CpG islands have been identified in 465 kb of DNA (Weber et al. 1991), whereas in a separate study of a 265-kb segment of DNA from a different region of 4p16.3, Carlock et al. (1992) have identified 13 transcriptional units. These gene frequencies, which were reported to be minimum estimates, would produce one gene approximately every 25 kb of DNA; this reflects a gene density similar to that seen in the MHC class III region. On the other hand, those regions of chromosomes that stain positively with Giemsa, the G-dark bands, may contain a much lower density of genes. An example of this is the G-dark band p21.3 on the X chromosome. This band contains the Duchenne muscular dystrophy gene, which spans ~2000 kb of DNA and lies in a region largely devoid of CpG islands (Burmeister et al. 1988). If one were to extrapolate to the rest of the human genome and argue that the gene density in G-light bands on average is one gene every 20 kb of DNA, then the G-light bands, which constitute approximately 42% of the chromosomal DNA, must contain about 70,000 genes. In addition, if this represented 74% of all the genes, then the human genome would contain about 100,000 genes. This is approximately double those estimates of 50,000–60,000 genes in the human genome, based on the frequency of CpG islands (Bickmore and Sumner 1989; Larsen et al. 1992).

The finding of such a large number of genes within the MHC strengthens the notion that there is a nonrandom distribution of genes in the human genome. Thus, with respect to the sequencing of the human genome, because of the considerable effort required, it may be more rewarding initially to direct those efforts toward sequencing the DNA corresponding to the G-light bands.

Acknowledgments

I thank Caroline Milner for helpful discussion, Ken Johnson for photographic help, and Alison Marsland for typing the manuscript. The work in the author's laboratory is supported by the Medical Research Council (UK) and the Arthritis and Rheumatism Council (UK).

References

Abe, K., J.F. Wei, F.S. Wei, Y.C. Hsu, H. Uehara, K. Artzt, and D. Bennett. 1988. Searching for coding sequences in the mammalian genome: The H-2K region of the mouse MHC is replete with genes expressed in embryos. *EMBO J.* **7:** 3441.

Adolph, S. and H. Hameister. 1990. In situ nick translation of human metaphase chromosomes with the restriction enzymes Msp l and Hpa II reveals an R-band pattern. *Cytogenet. Cell Genet.* **54:** 132.

Adweh, Z.L., D. Raum, E.J. Yunis, and C.A. Alper. 1983. Extended HLA/ complement allele haplotypes: Evidence for T/t-like complex in man. *Proc. Natl. Acad. Sci.* **80:** 259.

Allan, F.H. 1974. Linkage of HL-A and GBG. *Vox Sang.* **27:** 382.

Alper, C.A. 1981. Complement and the MHC. In *The role of the major*

histocompatibility complex in immunobiology (ed. M. Dorf), p. 173. Garland, New York.
Ashfield, R., P. Enriquez-Harris, and N.J. Proudfoot. 1991. Transcriptional termination between the closely-linked human complement genes C2 and Factor B: Common termination factor for C2 and c-myc? *EMBO J.* **10**: 4197.
Bach, F.H. and J.J. van Rood. 1976a. The major histocompatibility complex—Genetics and biology (part 1). *N. Engl. J. Med.* **295**: 806.
———. 1976b. The major histocompatibility complex—Genetics and biology (part 2). *N. Engl. J. Med.* **295**: 872.
———. 1976c. The major histocompatibility complex—Genetics and biology (part 3). *N. Engl. J. Med.* **295**: 927.
Bahram, S., D. Arnold, M. Bresnahan, J.L. Strominger, and T. Spies. 1991. Two putative subunits of a peptide pump encoded in the human major histocompatibility complex class II region. *Proc. Natl. Acad. Sci.* **88**: 10094.
Batchelor, J.R. and A.J. McMichael. 1987. Progress in understanding HLA and disease associations. *Br. Med. Bull.* **43**: 156.
Benacerraf, B. 1981. Role of MHC gene products in immune regulation. *Science* **212**: 1229.
Bickmore, W.A. and A.T. Sumner. 1989. Mammalian chromosome banding—An expression of genome organization. *Trends Genet.* **5**: 144.
Bird, A.P. 1987. CpG islands as gene markers in the vertebrate nucleus. *Trends Genet.* **3**: 342.
Bjorkman, P.J. and P. Parham. 1990. Structure, function and diversity of class I major histocompatibility complex molecules. *Annu. Rev. Immunol.* **59**: 253.
Blanck, G. and J.L. Strominger. 1988. Molecular organisation of the DQ subregion (DO-DX-DV-DQ) of the human MHC and its evolutionary implications. *J. Immunol.* **141**: 1734.
———. 1990. Cosmid clones in the HLA-DZ and -DP subregions. *Hum. Immunol.* **27**: 265.
Blank, V., P. Kourilsky, and A. Israel. 1992. NF-kB and related proteins: Rel/dorsal homologies meet ankyrin repeats. *Trends Biochem. Sci.* **17**: 135.
Bodmer, J.G., L.J. Kennedy, J. Lindsay, and A.M. Wasik. 1987. Applications of serology and the ethnic distribution of three locus HLA haplotypes. *Br. Med. Bull.* **43**: 94.
Bodmer, W.F., E. Albert, J.G. Bodmer, B. Dupont, B. Mach, S.G.E. Marsh, W.R. Mayr, P. Parham, T. Sasazuki, G.M.T. Schreuder, J. Strominger, A. Svejgaard, and P.I. Terasaki. 1988. Nomenclature of factors for the HLA system, 1991. *Immunogenetics* **36**: 135.
Bonetta, L., S.E. Kuehn, A. Huang, D.J. Law, L.M. Kalikin, M. Doi, A.E. Reeve, B.H. Brownstein, H. Yeger, B.R.G. Williams, and A.P.O. Feinberg. 1990. Wilms tumor locus on 11p13 defined by multiple CpG island-associated transcripts. *Science* **250**: 994.
Bossone, S.A., C. Asselin, A.J. Patel, and K.B. Marcu. 1992. MAZ, a zinc finger protein, binds to c-myc and C2 gene sequences regulating transcriptional initiation and termination. *Proc. Natl. Acad. Sci.* **89**: 7452.
Braciale, T.J. and V.L. Braciale. 1991. Antigen presentation: Structural themes and functional variation. *Immunol. Today* **12**: 124.
Bronson, S.K., J. Pei, P. Taillon-Miller, M.J. Chorney, D.E. Geraghty, and D.D.

Chaplin. 1991. Isolation and characterisation of YAC clones linking the HLA-B and HLA-C loci. *Proc. Natl. Acad. Sci.* **88**: 1676.

Burmeister, M., A.P. Monaco, E.F. Gillard, G.B. van Ommen, N.A. Affara, M.A. Ferguson-Smith, L.M. Kunkel, and H. Lehrach. 1988. A 10-megabase physical map of human Xp21, including the Duchenne muscular dystrophy gene. *Genomics* **2**: 189.

Campbell, R.D. and R.R. Porter. 1983. Molecular cloning and characterisation of the gene coding for human complement protein factor B. *Proc. Natl. Acad. Sci.* **80**: 4464.

Campbell, R.D., M.C. Carroll, and R.R. Porter. 1986. The molecular genetics of components of complement. *Adv. Immunol.* **38**: 203.

Campbell, R.D., I. Dunham, E. Kendall, and C.A. Sargent. 1990. Polymorphism of the human complement protein C4. *Exp. Clin. Immunogenet.* **7**: 69.

Carle, G.F. and M.V. Olson. 1984. Separation of chromosomal DNA molecules from yeast by orthogonal-field-alternation gel electrophoresis. *Nucleic Acids Res.* **12**: 5647.

Carlock, L., D. Wisniewski, A. Lorincz, A. Pandrangi, and T. Vo. 1992. An estimate of the number of genes in the Huntington disease gene region and the identification of 13 transcripts in the 4p16.3 segment. *Genomics* **13**: 1108.

Carroll, M.C., R.D. Campbell, and R.R. Porter. 1985a. Mapping of steroid 21-hydroxylase genes adjacent to complement component C4 genes in HLA, the major histocompatibility complex in man. *Proc. Natl. Acad. Sci.* **82**: 521.

Carroll, M.C., A. Palsdottir, K.T. Belt, and R.R. Porter. 1985b. Deletion of complement C4 and steroid 21-hydroxylase genes in the HLA class III region. *EMBO J.* **4**: 2547.

Carroll, M.C., R.D. Campbell, D.R. Bentley, and R.R. Porter. 1984. A molecular map of the human major histocompatibility complex class III region linking complement genes C4, C2 and factor B. *Nature* **307**: 237.

Carroll, M.C., P. Katzman, E.M. Alicot, B.H. Koller, D. Geraghty, H.T. Orr, J.L. Strominger, and T. Spies. 1987. A linkage map of the human major histocompatibility complex including the tumor necrosis factor genes. *Proc. Natl. Acad. Sci.* **84**: 8535.

Ceppellini, R., E.S. Curtoni, P.L. Mattiuz, V. Miggiano, G. Scudeller, and A. Serra. 1967. Genetics of leukocyte antigens: A family study of segregation and linkage. In *Histocompatibility testing* (ed. E.S. Curtoni et al.), p. 149. Munksgaard, Denmark.

Cerami, A. and B. Beutler. 1988. The role of cachectin/TNF in endotoxic shock and cachexia. *Immunol. Today* **9**: 28.

Chiang, H.L., S.R. Terlecky, C.P. Plant, and J.F. Dice. 1989. A role for a 70 kilodalton heat shock protein in lysosomal degradation of intracellular proteins. *Science* **246**: 382.

Chimini, G., J. Boretto, D. Marguet, F. Lanau, G. Lanquin, and P. Pontarotti. 1990. Molecular analysis of the human MHC class I region using yeast artificial chromosome clones. *Immunogenetics* **32**: 419.

Chimini, G., P. Pontarotti, C. Nguyen, A. Toubert, J. Boretto, and B.R. Jordan. 1988. The chromosome region containing the highly polymorphic HLA class I genes displays limited large scale variability in the human population. *EMBO J.* **7**: 395.

Collier, S., P.J. Sinnott, P.A. Dyer, D.A. Price, R. Harris, and T. Strachan. 1989. Pulsed field gel electrophoresis identifies a high degree of variability in the number of tandem 21-hydroxylase and complement C4 gene repeats in 21-hydroxylase deficiency haplotypes. *EMBO J.* **8:** 1393.

Colten, H.R. and S.B. Dowton. 1986. Regulation of complement gene expression. *Biochem. Soc. Symp.* **51:** 37.

Dausset, J. 1981. The major histocompatibility complex in man. *Science* **213:** 1469.

Davis, M.M. and P.J. Bjorkman. 1988. T-cell antigen receptor genes and T-cell recognition. *Nature* **334:** 395.

Dawkins, R.L., F.T. Christiansen, P.H. Kay, M. Garlepp, J. McCluskey, P.N. Hollingsworth, and P.J. Zilko. 1983. Disease associations with complotypes, supratypes and haplotypes. *Immunol. Rev.* **70:** 5.

DeMars, R. and T. Spies. 1992. New genes in the MHC that encode proteins for antigen processing. *Trends Cell Biol.* **2:** 81.

DeNagel, D.C. and S.K. Pierce. 1992. A case for chaperones in antigen processing. *Immunol. Today* **13:** 86.

Dice, J.F. 1990. Peptide sequences that target cytosolic proteins for lysosomal proteolysis. *Trends Biochem. Sci.* **15:** 305.

Dunham, I., C.A. Sargent, R.L. Dawkins, and R.D. Campbell. 1989a. An analysis of variation in the long-range genomic organization of the human major histocompatibility complex class II region by pulsed-field gel electrophoresis. *Genomics* **5:** 787.

———. 1989b. Direct observation of the gene organisation of the C4 and 21-hydroxylase loci by pulsed field gel electrophoresis. *J. Exp. Med.* **169:** 1803.

Dunham, I., C.A. Sargent, E. Kendall, and R.D. Campbell. 1990. Characterization of the class III region in different MHC haplotypes by pulsed-field gel electrophoresis. *Immunogenetics* **32:** 175.

Dunham, I., C.A. Sargent, J. Trowsdale, and R.D. Campbell. 1987. Molecular mapping of the human major histocompatibility complex by pulsed-field gel electrophoresis. *Proc. Natl. Acad. Sci.* **84:** 7237.

Dupont, B., S.E. Oberfield, E.M. Smithwick, T.D. Lee, and L.S. Levine. 1977. Close genetic linkage between HLA and congenital adrenal hyperplasia (21-hydroxylase deficiency). *Lancet* **II:** 1309.

Dupont, B., M.S. Pollack, L.S. Levine, G.J. O'Neill, B.R. Hawkins, and M.I. New. 1980. Congenital adrenal hyperplasia. Joint report from the 8th International Histocompatibility Workshop. In *Histocompatibility testing* (ed P. Terasaki), p. 693. UCLA Press, Los Angeles.

Duquesnoy, R.J. and M. Trucco. 1988. Genetic basis of cell surface polymorphisms encoded by the major histocompatibility complex in humans. *Crit. Rev. Immunol.* **8:** 103.

Flynn, G.C., T.G. Chappell, and J.E. Rothman. 1989. Peptide binding and release by proteins implicated as catalysts of protein assembly. *Science* **245:** 385.

Franke, U. and M.A. Pellegrino. 1977. Assignment of the major histocompatibility complex to a region on the short arm of human chromosome 6. *Proc. Natl. Acad. Sci.* **74:** 1147.

Freemont, P.S., I.M. Hanson, and J. Trowsdale. 1991. A novel cysteine-rich sequence motif. *Cell* **64:** 483.

Fu, S.M., H.G. Kunkel, H.P. Brusman, F.H. Allen, and M. Fotino. 1974. Evidence for linkage between HL-A histocompatibility genes and those involved in

the synthesis of the second component of complement. *J. Exp. Med.* **140:** 1108.
Gardiner, K., M. Horisberger, J. Kraus, U. Tantravahi, J. Korenberg, V. Rao, S. Reddy, and D. Patterson. 1990. Analysis of human chromosome 21: Correlation of physical and cytogenetic maps; gene and CpG island distributions. *EMBO J.* **9:** 25.
Gardiner-Garden, M. and M. Frommer. 1987. CpG islands in vertebrate genomes. *J. Mol. Biol.* **196:** 261.
Geraghty, D.E., B.H. Koller, and H.T. Orr. 1987. A human major histocompatibility complex class I gene that encodes a protein with a shortened cytoplasmic segment. *Proc. Natl. Acad. Sci.* **84:** 9145.
Geraghty, D.E., B.H. Koller, J.A. Hansen, and H.T. Orr. 1992a. The HLA class I gene family includes at least six genes and twelve pseudogenes and gene fragments. *J. Immunol.* **149:** 1934.
Geraghty, D.E., B.H. Koller, J. Pei, and J.A. Hansen. 1992b. Examination of four HLA class I pseudogenes: Common events in the evolution of HLA genes and pseudogenes. *J. Immunol.* **149:** 1947.
Geraghty, D.E., X. Wei, H.T. Orr, and B.H. Koller. 1990. HLA-F: An expressed HLA gene composed of a class I coding sequence linked to a novel transcribed repetitive element. *J. Exp. Med.* **171:** 1.
Geraghty, D.E., J. Pei, B. Lipsky, J.A. Hansen, P. Taillon-Miller, S.K. Bronson, and D.D. Chaplin. 1992c. Cloning and physical mapping of the HLA class I region spanning the HLA-E to HLA-F interval by using yeast artificial chromosomes. *Proc. Natl. Acad. Sci.* **89:** 2669.
Germino, G.G., D. Weinstat-Saslow, H. Himmelbauer, G.A.J. Gillespie, S. Somlo, B. Wirth, N. Barton, K.L. Harris, A.-M. Frischauf, and S.T. Reeders. 1992. The gene for autosomal dominant polycystic kidney disease lies in a 750 kb CpG-rich region. *Genomics* **13:** 144.
Gitelman, S.E., J. Bristow, and W.L. Miller. 1992. Mechanism and consequences of the duplication of the human C4/P450c21/gene X locus. *Mol. Cell. Biol.* **12:** 2124.
Glynne, R., S.H. Powis, S. Beck, A. Kelly, L.A. Kerr, and J. Trowsdale. 1991. A proteasome-related gene between two ABC transporter loci in the class II region of the human MHC. *Nature* **353:** 357.
Goldberg, A.L. and K.L. Rock. 1992. Proteolysis, proteasomes and antigen presentation. *Nature* **357:** 375.
Gorer, P.A. 1937. The genetic and antigenic basis of tumour transplantation. *J. Pathol. Bacteriol.* **44:** 691.
Hansen, J.A. and J.L. Nelson. 1990. Autoimmune diseases and HLA. *Crit. Rev. Immunol.* **10:** 307.
Hanson, I.M. and J. Trowsdale. 1991. Colinearity of novel genes in the class II regions of the MHC in mouse and human. *Immunogenetics* **34:** 5.
Hanson, I.M., A. Poustka, and J. Trowsdale. 1991. New genes in the class II region of the human major histocompatibility complex. *Genomics* **10:** 417.
Hanson, I.M., P. Gorman, V.C. Lui, K.S. Cheah, E. Solomon, and J. Trowsdale. 1989. The human alpha 2(XI) collagen gene (COL11A2) maps to the centromeric border of the major histocompatibility complex on chromosome 6. *Genomics* **5:** 925.
Hardy, D.A., J.I. Bell, E.O. Long, T. Liindsten, and H.O. McDevitt. 1986. Mapping of the class II region of the human major histocompatibility complex

by pulsed field gel electrophoresis. *Nature* **323**: 453.
Hauptmann, G., E. Grosshans, and E. Heid. 1974. Lupus érythémateux aigus et déficits héréditaires en complément. A propos d'un cas par déficit complet en C4. *Ann. Dermatol. Syphiligraphie* **101**: 479.
Hsieh, S.L. and R.D. Campbell. 1991. Evidence that gene G7a in human major histocompatibility complex encodes valyl-tRNA synthetase. *Biochem. J.* **278**: 809.
Inoko, H., K. Tsuji, V. Groves, and J. Trowsdale. 1989. Mapping of HLA class II genes by pulsed field gel electrophoresis and size polymorphism. In *Immunobiology of HLA* (ed. B. Dupont), vol. II, p. 83, Springer-Verlag, New York.
Kahloun, A.E., C. Vernet, A.M. Jouanolle, J. Boretto, V. Mauvieux, J.-Y. Le Gall, V. David, and P. Pontarotti. 1992. A continuous restriction map from HLA-E to HLA-F. Structural comparison between different HLA-A haplotypes. *Immunogenetics* **35**: 183.
Kelly, A.P., J.J. Monaco, S. Cho, and J. Trowsdale. 1991a. A new human HLA class II related locus, DM. *Nature* **353**: 571.
Kelly, A., S.H. Powis, R. Glynne, E. Radley, S. Beck, and J. Trowsdale. 1991b. Second proteasome-related gene in the human MHC class II region. *Nature* **353**: 667.
Kelly, A., S.H. Powis, L.A. Kerr, I. Mockridge, T. Elliott, J. Bastin, B. Uchanska-Ziegler, A. Ziegler, J. Trowsdale, and A. Townsend. 1992. Assembly and function of the two ABC transporter proteins encoded in the human major histocompatibility complex. *Nature* **355**: 641.
Kendall, E. 1992. "Molecular characterisation of the human major histocompatibility complex." Ph.D. thesis, Oxford University, England.
Kendall, E., C.A. Sargent, and R.D. Campbell. 1990. Human major histocompatibility complex contains a new cluster of genes between the HLA-D and complement C4 loci. *Nucleic Acids Res.* **18**: 7251.
Kendall, E., J.A. Todd, and R.D. Campbell. 1991. Molecular analysis of the MHC class II region in DR4, DR7, and DR9 haplotypes. *Immunogenetics* **34**: 349.
Khanna, A. and R.D. Campbell. 1993. Characterisation of a novel gene, G13, in the class III region of the human MHC. In *Proceedings of the 11th International Histocompatibility Workshop* (ed. K. Tsuji et al.), Oxford University Press, United Kingdom.
Klein, J. 1975. *Biology of the mouse histocompatibility-2 complex.* Springer-Verlag, New York.
Klein, J. and F. Figueroa. 1986. Evolution of the major histocompatibility complex. *Crit. Rev. Immunol.* **6**: 295.
Koller, B.H., D.E. Geraghty, Y. Shimizu, R. DeMars, and H.T. Orr. 1988. HLA-E: A novel HLA class I gene expressed in resting T-lymphocytes. *J. Immunol.* **141**: 897.
Koller, B.H., D.E. Geraghty, R. DeMars, L. Duvick, S.S. Rich, and H.T. Orr. 1989. Chromosomal organisation of the human MHC class I gene family. *J. Exp. Med.* **169**: 469.
Kostyu, D.D. 1991. The HLA gene complex and genetic susceptibility to disease. *Curr. Opin. Genet. Dev.* **1**: 40.
Kozono, H., S.K. Bronson, P. Taillon-Miller, M.K. Moorti, I. Jamry, and D.D. Chaplin. 1991. Molecular linkage of the HLA-DR, HLA-DQ, and HLA-DO genes in yeast artificial chromosomes. *Genomics* **11**: 577.

Lachmann, P.J., D. Grennan, A. Martin, and P. Demant. 1975. Identification of Ss protein as murine C4. *Nature* **258**: 242.

Langer, T. and W. Neupert. 1991. Heat shock proteins hsp60 and hsp70: Their roles in folding, assembly and membrane translocation of proteins. *Curr. Top. Microbiol. Immunol.* **167**: 3.

Larsen, F., G. Gundersen, R. Lopez, and H. Prydz. 1992. CpG islands as gene markers in the human genome. *Genomics* **13**: 1095.

Lawlor, D.A., J. Zemmour, P.D. Ennis, and P. Parham. 1990. Evolution of class-I MHC genes and proteins: From natural selection to thymic selection. *Annu. Rev. Immunol.* **8**: 23.

Lawrance, S.K. and C.L. Smith. 1990. Megabase scale restriction fragment length polymorphisms in the human major histocompatibility complex. *Genomics* **8**: 394.

Lawrance, S.K., C.L. Smith, R. Srivastava, C.R. Cantor, and S.M. Weissman. 1987. Megabase-scale mapping of the HLA gene complex by pulsed-field gel electrophoresis. *Science* **235**: 1387.

Levi-Strauss, M., M.C. Carroll, M. Steinmetz, and T. Meo. 1988. A previously undetected MHC gene with an unusual periodic structure. *Science* **240**: 201.

Lord, D.K., I. Dunham, R.D. Campbell, A. Bomford, T. Strachan, and T.M. Cox. 1990. Molecular analysis of the human MHC class I region in hereditary haemochromatosis—A study by pulsed-field gel electrophoresis. *Hum. Genet.* **85**: 531.

Lovett, M., J. Kere, and L.M. Hinton. 1991. Direct selection: A method for the isolation of cDNAs encoded by large genomic regions. *Proc. Natl. Acad. Sci.* **88**: 9628.

Maestrini, E., S. Rivella, C. Triboli, D. Purtilo, M. Rocchi, M. Archidiacono, and D. Toniolo. 1990. Probes for CpG islands on the distal long arm of the human X chromosome are clustered in Xq24 and Xq28. *Genomics* **8**: 664.

Matsumoto, K.I., N. Ishihara, A. Ando, H. Inoko, and T. Ikemura. 1992a. Extracellular matrix protein tenascin-like gene found in human MHC class III region. *Immunogenetics* **36**: 400.

Matsumoto, K.I., M. Arai, N. Ishihara, A. Ando, H. Inoko, and T. Ikemura. 1992b. Cluster of fibronectin type-III repeats found in the human major histocompatibility complex class III region shows the highest homology with the repeats in an extracellular matrix protein, tenascin. *Genomics* **12**: 485.

Mauff, G., M. Brenden, M. Braun-Stilwell, G. Doxiadis, C.M. Giles, G. Hauptmann, C. Rittner, P.M. Schneider, B. Stradmann-Bellinghausen, and B. Uring-Lambert. 1990. C4 reference typing report. *Complement Inflammation* **7**: 193.

McDevitt, H.O. and A. Chinitz. 1969. Genetic control of the antibody response: Relationship between immune response and histocompatibility (H-2) type. *Science* **163**: 1207.

McDevitt, H.O., B.D. Deak, D.C. Shreffler, J. Klein, J.H. Stimpfling, and G.D. Snell. 1972. Genetic control of the immune response: Mapping of the IR-1 locus. *J. Exp. Med.* **135**: 1259.

Meo, T., T. Krasteff, and D.C. Shreffler. 1975. Immunochemical characterization of murine H-2 controlled Ss (serum substance) protein through identification of its human homologues as the fourth component of complement. *Proc. Natl. Acad. Sci.* **72**: 4536.

Messer, G., J. Zemmour, H.T. Orr, P. Parham, E.H. Weiss, and J. Girdlestone. 1992. HLA-J, a second inactivated class I HLA gene related to HLA-G and HLA-A: Implications for the evolution of the HLA-A-related genes. *J. Immunol.* **148:** 4043.

Michaely, P. and V. Bennett. 1992. The ANK repeat: A ubiquitous motif involved in macromolecular recognition. *Trends Cell Biol.* **2:** 127.

Milner, C.M. 1991. "Characterisation of novel genes in the human major histocompatibility complex: The HSP70 and G9a genes." Ph.D. thesis, Oxford University, England.

Milner, C.M. and R.D. Campbell. 1990. Structure and expression of the three MHC-linked human HSP70 genes. *Immunogenetics* **32:** 242.

———. 1992. Polymorphic analysis of the three MHC-linked HSP70 genes. *Immunogenetics* **36:** 357.

———. 1993. The G9a gene in the human major histocompatibility complex encodes a novel protein containing ankyrin-like repeats. *Biochem. J.* (in press).

Monaco, A.P., R.L. Neve, C. Colletti-Feener, C.T. Bertelson, D.M. Kurnit, and L.M. Kunkel. 1986. Isolation of candidate cDNAs for portions of the Duchenne muscular dystrophy gene. *Nature* **323:** 646.

Morel, Y., J. Bristow, S.E. Gitelman, and W.L. Miller. 1989. Transcript encoded on the opposite strand of the human steroid 21-hydroxylase/complement component C4 gene locus. *Proc. Natl. Acad. Sci.* **86:** 6582.

Morgan, B.P and M.J. Walport. 1991. Complement deficiency and disease. *Immunol. Today* **12:** 301.

Okada, K., J.M. Boss, T. Spies, R. Mengler, C. Auffray, J. Lillie, D. Grossberger, and J.L. Strominger. 1985. Gene organisation of DC and DX subregions of the human major histocompatibility complex. *Proc. Natl. Acad. Sci.* **82:** 3410.

Olavesen, M.G., M. Snoek, and R.D. Campbell. 1993a. Localisation of a new gene adjacent to the HSP70 genes in the human and mouse MHCs. *Immunogenetics* (in press).

Olavesen, M.G., W. Thomson, J. Cheng, and R.D. Campbell. 1993b. Characterisation of a novel gene (G1) in the class III region of the human MHC. In *Proceedings of the 11th International Histocompatibility Workshop* (ed. K. Tsuji et al.). Oxford University Press, United Kingdom.

O'Neill, G.J., S.Y Yang, and B. Dupont. 1978. Two HLA-linked loci controlling the fourth component of human complement. *Proc. Natl. Acad. Sci.* **75:** 5165.

Orr, H.T. and R. DeMars. 1983. Class I-like HLA genes map telomeric to the HLA-A2 locus in human cells. *Nature* **302:** 534.

Partanen, J. and S. Koskimes. 1986. Human MHC class III genes, Bf and C4. Polymorphism, complotypes and association with MHC Class I genes in the Finnish population. *Hum. Hered.* **36:** 269.

Parham, P. 1990. Transporters of delight. *Nature* **348:** 674.

Parimoo, S., S.R. Patanjali, H.A. Shukla, D.D. Chaplin, and S.M. Weissman. 1991. cDNA selection: Efficient PCR approach for the selection of cDNAs encoded in large chromosomal DNA fragments. *Proc. Natl. Acad. Sci.* **88:** 9623.

Pontarotti, P., G. Chimini, C. Nguyen, J. Boretto, and B.R. Jordan. 1988. CpG islands and HTF islands in the HLA class I region: Investigation of the methylation status of class I genes leads to precise physical mapping of the HLA-B and -C genes. *Nucleic Acids Res.* **16:** 6767.

Prentice, H.L., P.M. Schneider, and J.L. Strominger. 1986. C4B gene polymorphism detected in a human cosmid clone. *Immunogenetics* **23:** 274.

Ragoussis, J., K. Bloemer, E.H. Weiss, and A. Ziegler. 1988. Localization of the genes for tumor necrosis factor and lymphotoxin between the HLA class I and III regions by field inversion gel electrophoresis. *Immunogenetics* **27:** 66.

Ragoussis, J., K. Bloemer, H. Pohla, G. Messer, E.H. Weiss, and A. Ziegler. 1989. A physical map including a new class I gene (cda12) of the human major histocompatibility complex (A2/B13 haplotype) derived from a monosomy 6 mutant cell line. *Genomics* **4:** 301.

Ragoussis, J., A. Monaco, I. Mockridge, E. Kendall, R.D. Campbell, and J. Trowsdale. 1991. Cloning of the HLA class II region in yeast artificial chromosomes. *Proc. Natl. Acad. Sci.* **88:** 3753.

Raum, D., Z. Awdeh, J. Anderson, L. Strong, J. Granados, L. Pevan, E. Giblett, E.J. Yunis, and C.A. Alper. 1984. Human C4 haplotypes with duplicated C4A or C4B. *Am. J. Hum. Genet.* **36:** 72.

Reddy, B.A., L.D. Etkin, and P.S. Freemont. 1992. A novel zinc finger coiled-coil domain in a family of nuclear proteins. *Trends Biochem. Sci.* **17:** 344.

Reinsmoen, N.L., P.S. Friend, W.V. Miller, A. Burgdorf, E.R. Gibbett, and E.J. Yunis. 1977. Inheritance of recombinant HLA-GLO haplotypes suggesting the gene sequence. *Nature* **267:** 276.

Rittner, C., C.M. Giles, M.L.H. Roos, P. Demant, and E. Mollenhauer. 1984. Genetics of human C4 polymorphism: Detection and segregation of rare and duplicated haplotypes. *Immunogenetics* **19:** 321.

Rollini, P., B. Mach, and J. Gorski. 1985. Linkage map of three HLA-DR β-chain genes: Evidence for a recent duplication event. *Proc. Natl. Acad. Sci.* **82:** 7197.

Sargent, C.A., I. Dunham, and R.D. Campbell. 1989a. Identification of multiple HTF-island associated genes in the human major histocompatibility complex class III region. *EMBO J.* **8:** 2305.

Sargent, C.A., I. Dunham, J. Trowsdale, and R.D. Campbell. 1989b. Human major histocompatibility complex contains genes for the major heat shock protein HSP70. *Proc. Natl. Acad. Sci.* **86:** 1968.

Schendel, D.J., G. O'Neill, and R. Wank. 1984. MHC-linked Class III genes. Analyses of C4 gene frequencies, complotypes and associations with distinct HLA haplotypes in German Caucasians. *Immunogenetics* **20:** 23.

Schmidt, C.M. and H.T. Orr. 1991. A physical linkage map of HLA-A, -G, -7.5p and -F. *Hum. Immunol.* **31:** 180.

Schwartz, D.C. and C.R. Cantor. 1984. Separation of yeast chromosome-sized DNAs by pulsed field gradient gel electrophoresis. *Cell* **37:** 67.

Servenius, B., K. Gustafsson, E. Widmark, E. Emmoth, G. Andersson, D. Larkhammer, L. Rask, and P. Peterson. 1984. Class II genes of the human major histocompatibility complex. Molecular map of the human HLA-SB (HLA-DP) region and sequence of an SBα (DPα) pseudogene. *EMBO J.* **3:** 3209.

Shaw, S., P. Kavathas, M.S. Pollack, D. Charmot, and C. Mawas. 1981. Family studies define a new histocompatibility locus, SB, between HLA-DR and GLO. *Nature* **293:** 745.

Shreffler, D.C. 1964. A serologically detected variant in mouse serum: Further evidence for genetic control by the Histocompatibility-2 locus. *Genetics* **49**: 973.

Shukla, H., G.A. Gillespie, R. Srivastava, F. Collins, and M.J. Chorney. 1991. A class I jumping clone places the HLA-G gene approximately 100kb from HLA-H within the HLA-A subregion of the MHC. *Genomics* **10**: 905.

Sinha, A.A., M.T. Lopez, and H.O. McDevitt. 1990. Autoimmune disease: The failure of self tolerance. *Science* **248**: 1380.

Snell, G.D. 1981. Studies in histocompatibility. *Science* **213**: 172.

Southern, E.M., R. Anand, W.R.A. Brown, and D.S. Fletcher. 1987. A model for separation of large DNA molecules by crossed field gel electrophoresis. *Nucleic Acids Res.* **15**: 5925.

Speiser, P.W. and P.C. White. 1989. Structure of the human RD gene: A highly conserved gene in the class III region of the major histocompatibility complex. *DNA* **8**: 745.

Spies, T. and R. DeMars. 1991. Restored expression of major histocompatibility class I molecules by gene transfer of a putative peptide transporter. *Nature* **351**: 323.

Spies, T., M. Bresnahan, and J.L. Strominger. 1989a. Human major histocompatibility complex contains a minimum of 19 genes between the complement cluster and HLA-B. *Proc. Natl. Acad. Sci.* **86**: 8955.

Spies, T., G. Blanck, M. Bresnahan, J. Sands, and J.L. Strominger. 1989b. A new cluster of genes within the human major histocompatibility complex. *Science* **243**: 214.

Spies, T., R. Sorrentino, J.M. Boss, K. Okada, and J.L. Strominger. 1985. Structural organisation of the DR subregion of the human major histocompatibility complex. *Proc. Natl. Acad. Sci.* **82**: 5165.

Spies, T., V. Cerundolo, M. Colonna, P. Cresswell, A. Townsend, and R. DeMars. 1992. Presentation of viral antigen by MHC class I molecules is dependent on a putative peptide transporter heterodimer. *Nature* **355**: 644.

Spies, T., M. Bresnahan, S. Bahram, D. Arnold, G. Blanck, E. Mellins, D. Pious, and R. DeMars. 1990. A gene in the human major histocompatibility complex class II region controlling the class I antigen presenting pathway. *Nature* **348**: 744.

Spring, B., C. Fonatsch, C. Muller, G. Pawelec, J. Kompf, P. Wernet, and A. Ziegler. 1985. Refinement of HLA gene mapping with induced B-cell mutants. *Immunogenetics* **21**: 277.

Strachan, T. 1987. Molecular genetics and polymorphism of class I HLA antigens. *Br. Med. Bull.* **43**: 1.

Tiwari, J.L. and P.I. Terasaki. 1985. *HLA and disease associations.* Springer-Verlag, New York.

Todd, J. 1990. Genetic control of autoimmunity in type 1 diabetes. *Immunol. Today* **11**: 122.

Tokunaga, K., G. Saueracker, P.H. Kay, F.T. Christiansen, R. Anand, and R.L. Dawkins. 1988. Extensive deletions and insertions in different MHC supratypes detected by pulsed field gel electrophoresis. *J. Exp. Med.* **168**: 933.

Tokunaga, K., P.H. Kay, F.T. Christiansen, G. Saueracker, and R.L. Dawkins. 1989. Comparative mapping of the human major histocompatibility complex in different racial groups by pulsed field gel electrophoresis. *Hum.*

Immunol. 26: 99.
Tokunaga, K., K. Omoto, T. Akaza, N. Akiyama, H. Ameniya, S. Naito, T. Sasazuki, H. Satoh, and T. Juji. 1985. Haplotype study of C4 polymorphism in Japanese. Associations with MHC alleles, complotypes, and HLA-complement haplotypes. *Immunogenetics* 22: 359.
Townsend, A. and H. Bodmer. 1989. Antigen recognition by class I-restricted T lymphocytes. *Annu. Rev. Immunol.* 7: 601.
Travers, P. and H.O. McDevitt. 1987. Molecular genetics of class II (Ia) antigens. In *The antigens* (ed. M. Sela), vol. VII, p. 147, Academic Press, New York.
Trowsdale, J. 1987. Genetics and polymorphism: Class II antigens. *Br. Med. Bull.* 43: 15.
Trowsdale, J., J. Ragoussis, and R.D. Campbell. 1991. Map of the human MHC. *Immunol. Today* 12: 443.
Trowsdale, J., I. Hanson, I. Mockridge, S. Beck, A. Townsend, and A. Kelly. 1990. Sequences encoded in the class II region of the MHC related to the "ABC" superfamily of transporters. *Nature* 348: 741.
Trowsdale, J., A. Kelly, J. Lee, S. Carson, P. Austin, and P. Travers. 1984. Linkage map of two HLA-SBβ and two HLA-SBα related genes: An intron in one of the SBβ genes contains a processed pseudogene. *Cell* 38: 241.
Uring-Lambert, B., J. Goetz, M.M. Tongio, S. Mayer, and G. Hauptmann. 1984. C4 haplotypes with duplications at the C4A or C4B loci: Frequency and association with Bf, C2 and HLA-A, B, C, DR alleles. *Tissue Antigens* 24: 70.
van Bleek, G.M. and S.G. Nathenson. 1992. Presentation of antigenic peptides by MHC class I molecules. *Trends Cell Biol.* 2: 202.
VanBuskirk, A.M., D.C. DeNagel, L.E. Guagliardi, F.M. Brodsky, and S.K. Pierce. 1991. Cellular and subcellular distribution of PBP72/74, a peptide-binding protein that plays a role in antigen processing. *J. Immunol.* 146: 500.
Venditti, C.P. and M.J. Chorney. 1992. Class I gene contraction within the HLA-A subregion of the human MHC. *J. Immunol.* (in press).
Weber, B., C. Collins, D. Kowbel, O. Riess, and M.R. Hayden. 1991. Identification of multiple CpG islands and associated conserved sequences in a candidate region for the Huntington disease gene. *Genomics* 11: 1113.
Wu, L.C., B.J. Morley, and R.D. Campbell. 1987. Cell-specific expression of the human complement protein factor B gene: Evidence for the role of two distinct 5' flanking elements. *Cell* 48: 331.
Yu, C.Y. 1991. The complete exon-intron structure of a human complement component C4A gene. DNA sequences, polymorphism, and linkage to the 21-hydroxylase gene. *J. Immunol.* 146: 1057.
Yu, C.Y., K.T. Belt, C.M. Giles, R.D. Campbell, and R.R. Porter. 1986. Structural basis of the polymorphism of human complement components C4A and C4B: Gene size, reactivity and antigenicity. *EMBO J.* 5: 2873.
Yunis, E.J. and D.B. Amos. 1971. Three closely linked genetic systems relevant to transplantation. *Proc. Natl. Acad. Sci.* 68: 3031.
Zhang, W.J., M.A. Degli-Esposti, T.J. Cobain, P.U. Cameron, F.T. Christiansen, and R.L. Dawkins. 1990. Differences in gene copy number carried by different MHC ancestral haplotypes. *J. Exp. Med.* 171: 2101.
Ziegler, A., L.L. Field, and A.Y. Sakaguchi. 1991. Report of the committee on the genetic constitution of chromosome 6. *Cytogenet. Cell Genet.* 58: 295.

Molecular Genetics and Physical Mapping in Human Xp21

Anthony P. Monaco, Ann P. Walker, Meng F. Ho, Jamel Chelly, Edward Clarke, Yumiko Ishikawa-Brush, and Françoise Muscatelli

The Human Genetics Laboratory, Imperial Cancer Research Fund
Institute of Molecular Medicine, John Radcliffe Hospital
Headington, Oxford, OX3 9DU, United Kingdom

This chapter focuses on the present state of the physical map of the human cytogenetic band Xp21 with respect to gene loci, anonymous DNA probes, long-range pulsed field gel electrophoresis (PFGE) maps, and yeast artificial chromosome (YAC) contigs. The current knowledge concerning the position of disease gene loci is discussed, as well as progress in isolating corresponding expressed sequences by positional cloning techniques.

The main topics discussed include:

❑ introduction and overview of Xp21 contiguous deletion syndromes and physical mapping and positional cloning of disease genes (DMD and CYBB)

❑ the Duchenne (DMD) and Becker (BMD) muscular dystrophy gene, including a physical map of the locus constructed by PFG mapping, YAC contigs, a review of 5′ and 3′ gene promoters, and advances in methods for detecting deletions and point mutations

❑ the glycerol kinase deficiency (GK), congenital adrenal hypoplasia (AHC), and Oregon eye disease (OED) gene loci, including an overview of the physical map distal to DMD and critical regions for AHC and GK gene loci, based on analysis of patient DNA, deletion junction cloning, PFG mapping, and construction of YAC contigs

❑ discussion of the physical map and YAC contigs for the region in proximal Xp21 containing McLeod syndrome (XK), chronic granulomatous disease (CYBB), retinitis pigmentosa form 3 (RP3), and ornithine transcarbamylase deficiency (OTC) gene loci, as well as progress toward isolating the disease genes responsible for XK and RP3 by positional cloning techniques

INTRODUCTION

Xp21 contiguous deletion syndromes

The cytogenetic band Xp21 on the short arm of the human X chromosome has been one of the most intensively studied areas of the human genome because of the disease gene loci located in this region. Historically, most of this work centered on the hunt for the DMD gene because of its high frequency in the population. This work revealed that DMD and other disease gene loci in Xp21 are associated with deletions of the chromosome region. When a disease phenotype such as DMD is exhibited in the same patient with one or more clinically distinct disease phenotypes, large deletions are found with two or more genes being absent. The phenomenon of deletions giving rise to more than one phenotype was originally termed contiguous gene syndrome (Schmickel 1986) and more recently, contiguous deletion syndrome (Ballabio 1991). Large deletions in the Xp21 region have been useful for the isolation of genes for DMD and CYBB. Overlapping deletions found in patients with different combinations of disease phenotypes have also been essential in ordering and localizing other disease loci in Xp21. Besides Xp21, contiguous deletion syndromes have been found in other X chromosome regions and were important in positioning and/or isolating genes for steroid sulfatase deficiency and Kallmann syndrome in Xp22.3; choroideremia, deafness, and mental retardation in Xq21; and in various autosomal loci (for review, see Ballabio 1991).

The first male patient studied who exhibited a contiguous deletion syndrome in Xp21 was BB, who had four X-linked diseases: DMD, CYBB, XK, and RP3. Cytogenetic analysis of chromosomes from the patient BB revealed a visible interstitial deletion in part of the Xp21 region, and by hybridization analysis of DNA immobilized on membranes, a DNA probe (p754, DXS84) was found to be deleted (Francke et al. 1985). Many other patients exhibiting different combinations of diseases and deletion syndromes were then studied at the cytogenetic and molecular level, and this gave rise to the following gene order from telomere to centromere in Xp21: ZFX - POLA - AHC - GK - DMD - XK - CYBB - RP3 - OTC (Fig. 1). The physical mapping experiments that gave this order of genes in Xp21 are discussed in individual sections.

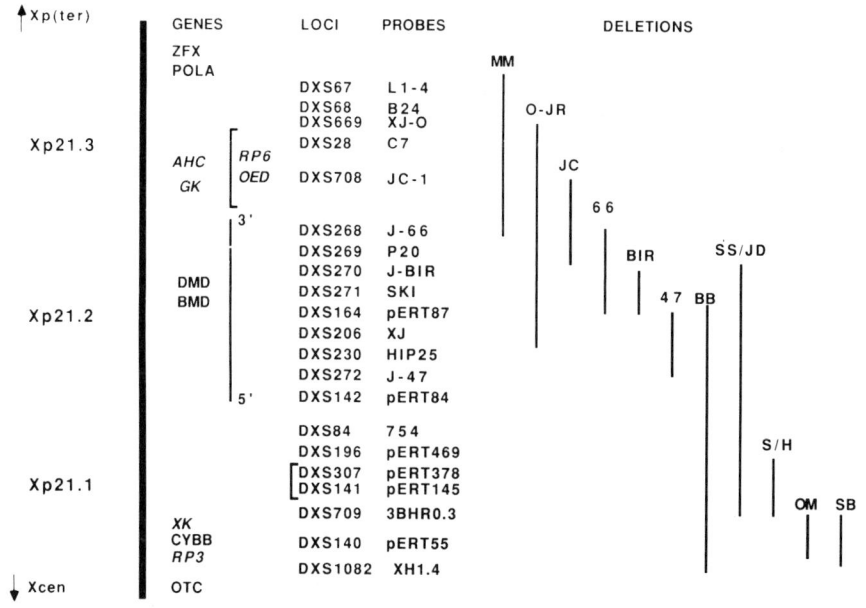

	GENES	LOCI	PROBES	DELETIONS
↑Xp(ter)	ZFX			
	POLA			MM
		DXS67	L1-4	
		DXS68	B24	O-JR
		DXS669	XJ-O	
Xp21.3		DXS28	C7	
	AHC RP6			JC
	GK OED	DXS708	JC-1	
				66
		3' DXS268	J-66	
		DXS269	P20	
		DXS270	J-BIR	BIR SS/JD
	DMD	DXS271	SKI	
Xp21.2	BMD	DXS164	pERT87	47 BB
		DXS206	XJ	
		DXS230	HIP25	
		DXS272	J-47	
		5' DXS142	pERT84	
		DXS84	754	
		DXS196	pERT469	S/H
Xp21.1		⌈DXS307	pERT378	
		⌊DXS141	pERT145	
		DXS709	3BHR0.3	OM SB
	XK			
	CYBB	DXS140	pERT55	
	RP3	DXS1082	XH1.4	
↓Xcen	OTC			

Figure 1 Schematic diagram of Xp21 with the order of genes, loci, and probes indicated from Xpter to Xcen. ZFX and POLA are abbreviations for the X-linked zinc finger gene and DNA polymerase α, respectively. Other gene abbreviations are described in the text. Uncloned genes are in italics and a bracket indicates order unknown. On the right are shown several of the deletions in patients which helped define the order of genes and probes. Several of the deletion breakpoints have been isolated from the DNA of the patients (O-JR, JC, 66, BIR, 47, BB, and JD).

Isolation of the DMD and CYBB genes

The molecular analysis of the deletion found in patient BB was considerably furthered by the development of differential hybridization techniques that enriched for DNA fragments missing from this region of Xp21 (phenol emulsion reassociation technique [pERT] probes) (Kunkel et al. 1985). The resulting DNA fragments were then used as hybridization probes to detect deletions in DNA isolated from patients exhibiting only the DMD phenotype, thereby pinpointing which DNA fragment (pERT87, DXS164) was contained within the DMD gene (Monaco et al. 1985). Chromosome walking from pERT87 and isolation of single copy DNA fragments which were hybridized to DNA from different species ("zoo blots") identified conserved fragments and led to the detection of expressed portions of the DMD gene (14 kb) that seemed to be spread across a large section of Xp21 (Monaco et al. 1986). In parallel experiments, a balanced X;21 translocation in a female patient with DMD was used to isolate DNA fragments from within the DMD gene (Ray et al. 1985). A DNA probe from the chromosome 21 portion of the translocation breakpoint was used to isolate a fragment from patient DNA that contained the translocation breakpoint and, therefore, DNA from both chromosome 21 and Xp21 (pXJ, DXS206). The Xp21-specific XJ probe was used for chromosome walking and resulted in independent isolation of expressed portions of the large DMD gene (Burghes et al. 1987).

Analysis of DNA from patient NF (Kousseff 1981; Baehner et al. 1986), who exhibited DMD, CYBB, and XK, revealed a deletion that was similar in size to the BB deletion (at least missing all the same pERT probes and DXS84). RNA isolated from a cell line derived from NF was used in a subtractive hybridization protocol to isolate the gene responsible for chronic granulomatous disease (CYBB; Royer-Pokora et al. 1986). The differential hybridization experiment enriched for cDNAs that were deleted in the patient and derived from induced granulocytic HL60 cells. The enriched cDNA population was hybridized to genomic DNA clones known to be deleted in DNA isolated from patients NF and BB, in order to identify an expressed DNA fragment. This DNA fragment was then used to probe a cDNA library constructed from RNA from the induced HL60 cells. A cDNA clone was isolated and found to be rearranged in several patients exhibiting CYBB only, thereby indicating that it was most likely the gene responsible for chronic granulomatous disease.

Positional cloning and new technologies

Since the isolation of the genes for CYBB, DMD, and other diseases by what is now termed "positional cloning" (Collins 1992), new technologies have emerged that have made the isolation of disease genes more efficient. One important advance has been the ability to isolate large DNA fragments (50–1500 kb) as linear artificial chromosomes in yeast (YACs, Burke et al. 1987). This technique has enhanced the speed of chromosome walking toward regions containing disease genes, and cloned DNA sequences are better represented in YAC libraries than in prokaryotic-based genomic libraries (Coulson et al. 1988). As discussed in the following sections, the complete DMD gene has now been isolated and reconstructed using YACs, and contiguous sets of overlapping YAC clones (contigs) have been obtained containing the genes for AHC and GK distal to DMD; and XK, CYBB, RP3, and OTC proximal to DMD.

The second advance has been the development of new methods to find expressed sequences in genomic DNA, other than detecting conservation of sequences on zoo blots, differential hybridization strategies, and identification of CpG islands (Lindsay and Bird, 1987). Whole YAC inserts and cosmid clones have been used to directly isolate cDNA clones by hybridization to libraries, after effectively depleting the probe of repeat sequences by preannealing to total human or mouse DNA (Elvin et al. 1990; Herrmann et al. 1990). Exon amplification is a new method that utilizes exon:intron splicing sequences in genomic DNA to trap exons after transfection of DNA fragments ligated in the appropriate vector into mammalian cells (Duyk et al. 1990; Buckler et al. 1991). Reverse transcriptase polymerase chain reaction (RT-PCR) amplification of RNA results in PCR products containing the spliced exons. The resulting exons can be subcloned, sequenced, and used as hybridization probes on cDNA libraries. Alternatively, specific oligonucleotide primers based on the exon sequence can be used to extend to both the 5' and 3' ends of RNA by PCR methods known as rapid amplification of cDNA ends (RACE; Frohman et al. 1988; Loh et al. 1989). The use of cDNA selection procedures also offers promise to isolate genes (Lovett et al. 1991; Parimoo et al. 1991). This method involves hybridization of PCR-amplified cDNA library inserts to genomic DNA fragments from YAC and cosmid contigs after repetitive sequences are effectively blocked by competitor DNA in both the probe and target DNA. The hybridization is followed by stringent washing, then elution of specifically hybridized cDNA inserts. Further enrichment of the cDNA inserts specific to the genomic DNA of interest can be achieved by successive PCR and hybridization cycles. Recently, such cDNA selection methods have been applied to contigs of cloned genomic DNA from the Xq28 region with the successful enrichment of many genes from the region (Korn et al. 1992). This method also offers the possibility to test cDNA libraries from several different tissues, developmental stages, and cell lines in parallel, thereby allowing a more efficient and comprehensive screening procedure for all genes from a certain region.

With the development of new sequencing technologies as the genome project advances, it will become easier to determine the complete nucleotide sequence of a critical region known to contain a disease gene. By using algorithms to search for exon sequences in the complete genomic DNA sequence (Uberbacher and Mural 1991), the amino acid coding regions will be easier to identify. This has already been demonstrated by sequencing 67 kb in Xp22.3 that resulted in the identification of coding sequences for the Kallmann syndrome gene (Legouis et al. 1991). It will also be interesting to sequence some of the large introns in the DMD gene to look for new promoters or regulatory elements, or other completely separate transcription units.

THE DMD GENE

Clinical phenotype and dystrophin

DMD is a severe muscle-wasting disease that affects about 1 of 3500 live male births (Moser 1984). Patients usually lose the ability to walk by the age of 12 years and are confined to wheelchairs. Death ensues by the late teens or mid-20s from respiratory or cardiac insufficiency. BMD is a less severe form of X-linked muscular dystrophy that affects about 1 of 30,000 males. In contrast to DMD, the onset and clinical severity of muscle weakness in patients with BMD tends to be more heterogeneous. About one-third of X-linked muscular dystrophy patients are mentally retarded.

The gene responsible for both DMD and BMD encodes a 14-kb mRNA and a 427-kD protein called dystrophin that is localized to the muscle cytoskeleton (for review, see Monaco 1989). From the predicted amino acid sequence, dystrophin is a rod-shaped structural protein with an amino-terminal actin-binding domain and long spectrin-like triple helical repeats along its length that may contain hinge segments for flexibility (Koenig et al. 1988; Koenig and Kunkel 1990). Recent data have shown that dystrophin binds to a complex of integral membrane glycoproteins most likely near its carboxy-terminal end (Ervasti and Campbell 1991). One of these membrane proteins also binds laminin, thus suggesting that the extracellular matrix is linked to dystrophin via this complex of membrane glycoproteins (Ibraghimov-Beskrovnaya et al. 1992).

Gene organization and physical maps

From physical mapping of the heterogeneous locations of deletion and translocation breakpoints (Kunkel et al. 1986; Boyd et al. 1987) and the genomic organization of small exons and large introns (Monaco et al. 1986), it was predicted that the DMD gene could be 1-2 megabase pairs (Mbp) in size. The large size (2300 kb) was confirmed by long-range mapping using PFGE and hybridization analysis of genomic DNA fragments obtained from chromosome walking and deletion jumping experiments (Burmeister and Lehrach 1986; Van Ommen et al. 1986, 1987; Kenwrick et al. 1987; Burmeister et al. 1988). On the basis of its complete cDNA sequence, the DMD gene was predicted to have a minimum of 65 exons (Koenig et al. 1987, 1988) and has recently been determined to have 79 exons (Roberts et al. 1992b). The extent of deletions and duplications occurring within the gene was determined by hybridizing the cDNA clones to patient DNA on Southern blots. The results from several studies indicated that about 65% of mutations were deletions and 5% were duplications; a "hot spot" for breakpoints occurred in the middle of the gene, with few deletions being located at the 3' end of the gene (Den Dunnen et al. 1987; Forrest et al. 1987; Koenig et al. 1987; Darras et al. 1988; Wapenaar et al. 1988; Hu et al. 1990). The PFGE map of *Sfi*I sites along the gene was used for diagnostic purposes to determine the extent of deletions and duplications found in patient DNA and, more importantly, to determine carrier status (Den Dunnen et al. 1987, 1989). The *Sfi*I PFGE map was also used to localize the translocation breakpoints giving rise to DMD and BMD in females (Meitinger et al. 1988). The severity of the DMD versus BMD phenotype was shown in most cases to depend on the effect of intragenic deletions or duplications on the open reading frame of the spliced transcript (Malhotra et al. 1988; Monaco et al. 1988; Baumbach et al. 1989; Koenig et al. 1989; Hu et al. 1990). Exceptions to the open reading frame hypothesis may be due to alternatively spliced transcripts (Chelly et al. 1990).

Dystrophin gene diversity

With the identification of a human brain-specific promoter (Feener et al. 1989) located about 100 kb proximal to the muscle-specific promoter (Boyce et al. 1991), the estimated size of the DMD gene determined by PFGE is about 2400 kb. More recently, several new promoters and transcriptional start sites have been identified that are utilized in different tissues (Fig. 2). Gòrecki et al. (1992) have identified two new alternative dystrophin transcripts in Purkinje cells of the cerebellum and in the dentate gyrus of the hippocampus by in situ hybridization of radiolabeled exon-specific oligonucleotides on mouse brain sections. They isolated and sequenced both the human and mouse Purkinje cell-specific first exon and untranslated region and mapped it in human genomic DNA about midway between the muscle-specific promoter and the common exon 2 in a large 240-kb intron. The dentate gyrus transcript seems to be produced from a more distal location and is currently under investigation to determine its precise origin in the DMD gene.

At the 3'end of the DMD gene, a novel transcript of 6.5 kb was identified and found to be present in tissues much different from the normal dystrophin isoforms (Bar et al. 1990). Three groups have independently cloned and sequenced a new exon and potential promoter that maps in genomic DNA between exons 62 and 63 (Blake et al. 1992; Hugnot et al. 1992; Lederfein et al. 1992; Rapaport et al. 1992). The protein product was found to be approximately 70–80 kD and was abundant in many tissues except adult skeletal muscle. Further investigation is required to determine whether this new exon gives rise to the original 6.5-kb transcript reported by Bar et al. (1990) or a 4.8-kb transcript abundant in Schwannoma cells (Blake et al. 1992). The diversity of the dystrophin gene transcripts is becoming increasingly complex, with a minimum of four promoters and the previously reported alternative splicing near the 3'end of the gene in both human and mouse (Feener et al. 1989; Bies et al. 1992).

DMD YAC contigs and reconstruction

Although much effort had been invested to isolate the entire DMD 2400-kb genomic locus in bacteriophage and cosmid clones, only about one-quarter of the gene was recovered in cloned DNA fragments because of the small insert size (Monaco et al. 1986, 1987; Burghes et al. 1987;

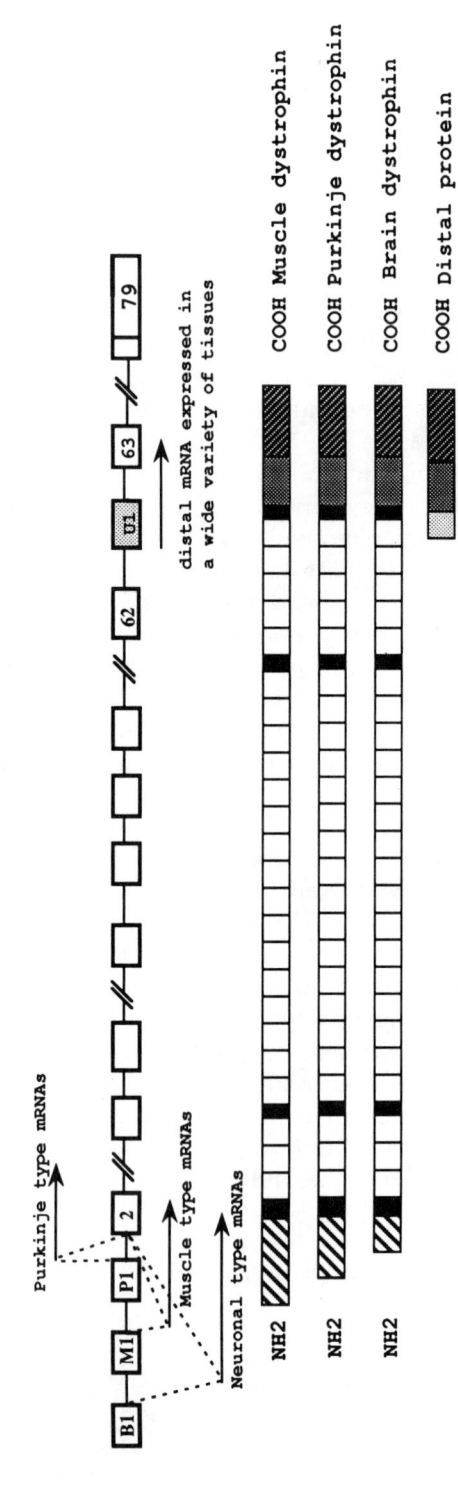

Figure 2 Diagram showing the diversity of dystrophin gene transcripts. There are three promoters isolated from the 5' end upstream of the common exon 2, giving rise to brain-type (B1), muscle-type (M1), and Purkinje cell-type (P1) transcripts. There is also a transcript from a promoter (U1) between exons 62 and 63 which is expressed in a variety of non-muscle tissues.

Heilig et al. 1987; Van Ommen et al. 1987; Blonden et al. 1989). With the introduction of YACs, capable of containing large inserts, the isolation of the complete DMD locus became more feasible. Two groups have constructed overlapping YAC contigs covering the entire DMD gene, using either a colony hybridization-based screening strategy (Fig. 3) (Monaco et al. 1992) or a PCR-based screening strategy with individual probes identified as sequence-tagged sites (STS) in DNA (Olson et al. 1989; Coffey et al. 1992). Both contigs showed the ability of YACs to cover a large gene with no apparent gaps or "unclonable" regions.

The YAC contigs were used to confirm and extend the *Sfi*I long-range PFGE map (Fig. 3) and helped to order and position several exons near the 3' end of the gene. These YAC clones have also proved useful in mapping the Purkinje cell-specific exon (Gòrecki et al. 1992) and in isolating and sequencing exons at the 3' end of the gene (Roberts et al. 1992b).

YACs from these two DMD contigs were reconstructed into one YAC using homologous recombination during meiosis in yeast. The resulting 2.4-Mbp YAC contains the complete DMD gene minus exon 60, and the brain-specific promoter and first exon (Den Dunnen et al. 1992). There seem to be sequences around exon 60 that become unstable during or after meiotic recombination. With the addition of the brain-specific promoter region and the introduction of exon 60, it will soon be possible for the complete DMD gene to be represented on one stable YAC with all the coding and regulatory sequences necessary for its complex diversity of transcripts. Transfer of this YAC by spheroplast fusion to mammalian cells in culture (for review, see Huxley and Gnirke 1991), or to embryonic stem cells to generate transgenic mice, will allow experiments to study the regulation of transcription from different promoters and alternatively spliced exons for this giant gene.

Detection of DMD mutations

As described previously, about 70% of mutations within the dystrophin gene are due to gross rearrangements such as deletions, duplications, and translocations. Mapping the deletion and duplication breakpoints in patient DNA has allowed accurate prenatal diagnosis and carrier detection to be performed. Many of the exon:intron borders of the gene have been sequenced, and strategies have been developed to detect deletions using different intron sequences as primers in PCR-based assays. It is now possible to detect 98% of deletions by multiplex PCR amplification (Chamberlain et al. 1988, 1990; Beggs et al. 1990). Recently, amplification of the dystrophin gene by RT-PCR of RNA derived from any tissue or cell line (Chelly et al. 1989) has allowed the study of deletions at the level of the transcripts produced. The results show that there is a series of alternatively spliced transcripts which are specific to the deleted geno-

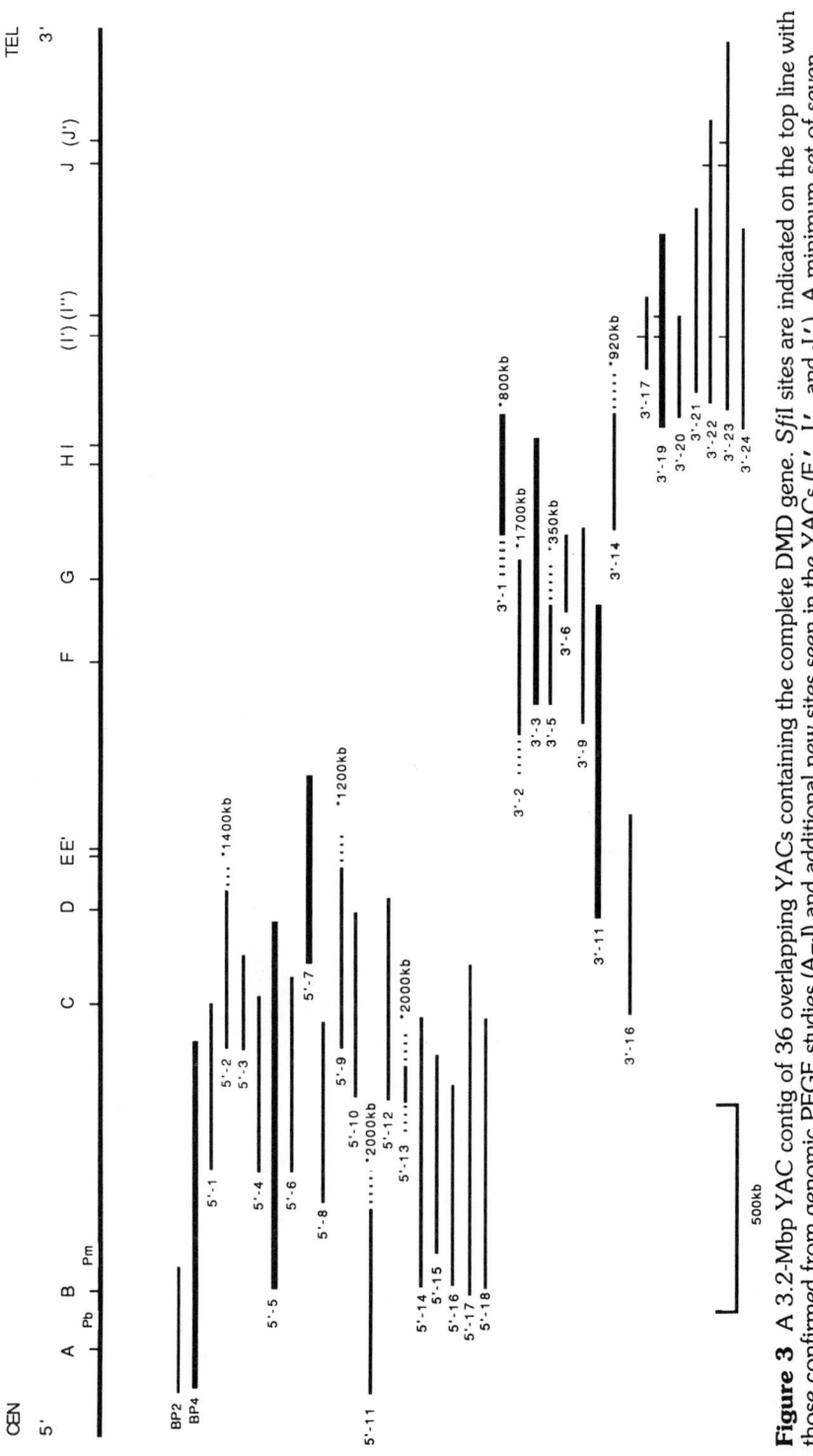

Figure 3 A 3.2-Mbp YAC contig of 36 overlapping YACs containing the complete DMD gene. *Sfi*I sites are indicated on the top line with those confirmed from genomic PFGE studies (A–J) and additional new sites seen in the YACs (E', I', and J'). A minimum set of seven YACs that contain the entire 2.4-Mbp DMD gene are indicated as bold lines; aberrant or chimeric sections of YACs are shown as dashed lines with an asterisk next to the YAC size in kb. YAC 3'-23 also contains the DXS708 locus, the GK locus, and at least part of the AHC gene (Walker et al. 1992). (Reprinted, with permission, from Monaco et al. 1992.)

types. These transcripts can shift the reading frame compared to the original deletion and may explain exceptions to the open reading frame hypothesis of DMD versus BMD genotypes and phenotypes (Chelly et al. 1990; Roberts et al. 1991).

Of all dystrophin gene mutations, however, 30% are not so easily detectable even by PCR methods and are most likely point mutations or small deletions. This represents a significant diagnostic problem because of the high rate of recombination across the DMD gene (up to 12% of meioses; Abbs et al. 1990), making diagnosis based only on genetic markers inaccurate. Using RT-PCR of RNA coupled with chemical mismatch detection (Cotton et al. 1988), point mutations were detected in seven DMD patients with no apparent deletions (Roberts et al. 1992a). Most of the point mutations were nonsense mutations causing premature termination of translation and truncation of the dystrophin product. The nonsense mutations, in effect, lead to the same truncated dystrophin products as deletions that disrupt the open reading frame of the transcript and result in a downstream nonsense codon. However, even with this technology, detecting point mutations routinely in such a large gene will not be an easy diagnostic test.

AHC AND GK DEFICIENCY LOCI
Clinical phenotypes

Patients with contiguous deletion syndromes in distal Xp21 can exhibit various combinations of diseases such as AHC, GK, DMD, mental retardation and a specific eye disease now called Oregon eye disease (OED). AHC is a severe disease with almost complete absence of adrenal gland function, known as primary adrenal insufficiency. Patients require replacement therapy with glucocorticoids, mineralocorticoids, and salt for survival. The growth of AHC patients is retarded, and during adolescence they fail to mature sexually due to gonadotrophin deficiency leading to hypogonadism. Patients with GK deficiency have elevated levels of glycerol in the blood and urine and can be diagnosed incorrectly sometimes as hypertriglyceridemic (McCabe 1989). Mental retardation and OED are other components of this contiguous deletion syndrome but may in fact be secondary to dystrophin deficiency, although this is still under investigation.

OED was detected in a patient with AHC, GK, DMD, and a large deletion of the region distal to DMD and was previously thought to be a clinically similar eye disease found in the Åland islands (Åland island eye disease, AIED; Weleber et al. 1989; Pillers et al. 1990). However, when the original AIED was mapped by genetic linkage analysis to the pericentromeric region (Alitalo et al. 1991), the eye disease associated with distal Xp21 deletions was renamed OED (Davies et al. 1991) after the origin of the patient in which it was described. The distal deletion breakpoint was isolated from analysis of patient DNA (O-JR; Fig. 1) and mapped between DXS68 and DXS28 (XJ-0, DXS669; Pillers et al. 1990). Another eye disease mapping in the region distal to DMD is retinitis pigmentosa form 6 (RP6) positioned here by genetic linkage analysis in one family (Musarella et al. 1988, 1990; Ott et al. 1990).

Physical mapping and gene localization distal to DMD

Cytogenetic and molecular analysis of deletions found in the DNA of patients exhibiting combinations of AHC, GK, and DMD positioned the AHC and GK genes between the 3'end of DMD and a distal group of anonymous DNA probes L1-4 (DXS68), B24 (DXS67), and C7 (DXS28) (Figs. 1 and 4) (Patil et al. 1985; Weiringa et al. 1985; Bartley et al. 1986; Dunger et al. 1986; Wilcox et al. 1986; Francke et al. 1987; Yates et al. 1987; Chelly et al. 1988; Darras and Francke 1988; Davies et al. 1988, 1991; McCabe et al. 1989; Towbin et al. 1990). The distal group of three anonymous markers had been published in five different possible orders, but new data from fluorescence in situ mapping in interphase nuclei (Trask et al. 1992), YAC contig analysis (Fig. 4) (Walker et al. 1991), and patient deletion mapping (Bartley and Gies, 1989) favor a sixth (and, we hope, final) order as pter-DXS68-DXS67-DXS28-cen.

The first published PFGE map of the region between the 3'end of DMD and this group of three probes (Burmeister et al. 1988) indicated the existence of a large region with few CpG islands. These data were confirmed by Love et al. (1990), who isolated a deletion junction fragment from a patient (JC) exhibiting DMD and GK only. The JC deletion junction fragment (JC-1, DXS708) was therefore positioned somewhere between the AHC and GK gene loci. When JC-1 was hybridized to DNA partially digested with *Bss*HII and resolved by PFGE up to 4.2 Mbp, it could not be linked to DMD or DXS28. Thus, the minimum size of the region thought to contain the AHC and GK genes was more than 4 Mbp.

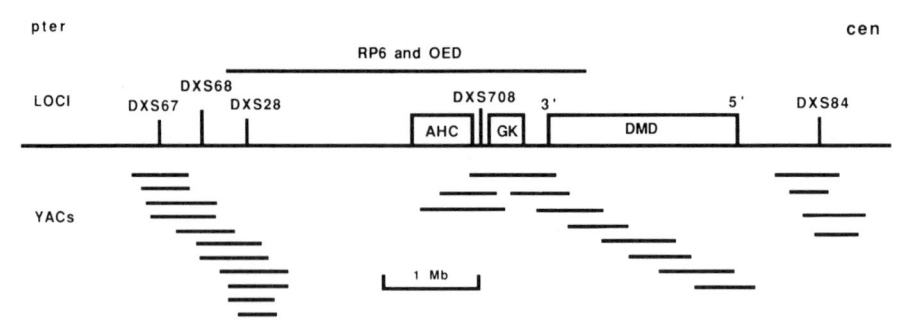

Figure 4 Diagram indicating the positions of genes, anonymous DNA probes, and YAC contigs in distal Xp21. The YAC contig covering DXS67, DXS68, and DXS28 confirmed the controversial order of these three probes (Walker et al. 1991). The YAC contig extending distally from the 3'end of DMD contains the GK gene and at least part of the AHC locus (Walker et al. 1992).

More recently, two DNA markers, DXS727 and DXS319, have been mapped to this large interval between DMD and DXS28 (Worley et al.1992). DXS727, DXS319, and another probe within 10 kb of the 3'end of DMD (DXS726), were sequenced, and STS primers were described for easy analysis of patient DNA samples by PCR. Both DXS727 and DXS319 were found to be deleted in one patient with AHC only, indicating their location distal to the GK locus and near the AHC locus. From an additional patient with AHC, GK, and DMD who had DXS28 and DXS727 present, but DXS319 absent, the order could be determined for these probes as pter-DXS28-DXS727-DXS319-DXS726-DMD-cen.

YAC contigs and critical regions for AHC and GK

Hybridization of the deletion junction clone JC-1 (DXS708) to the ICRF YAC library (Larin et al. 1991) identified several YACs, one of which was also positive for the 3'end of DMD. This YAC clone was 890 kb and is indicated in Figure 3 as DMD 3'-23. Partial digest mapping of DMD YACs 3'-23 and 3'-19, and two additional YACs isolated with DXS708, enabled a 1.2-Mbp map to be constructed extending distally from the 3' end of DMD (Fig. 4) (Walker et al. 1992). DMD 3'-23 was hybridized directly to the ICRF X chromosome cosmid library (Nizetic et al. 1991), and 39 cosmids were isolated. Single copy fragments isolated from 10 cosmids were used for mapping the YAC contig, identifying overlaps between cosmid clones, and analyzing the extent of deletions in patient DNA. Thirteen single copy probes, plus DXS28 and DXS708, were hybridized to DNA from 20 patients with various combinations of AHC, GK, and DMD (Walker et al. 1992). The results indicated that the GK gene could be localized to a deletion interval region of 50–250 kb in proximity to DXS708 (Fig. 4). Nonoverlapping deletions found in DNA from AHC patients suggested that the AHC gene may be large and may extend over at least 200–500 kb of the YAC contig. These new data provide a more accurate localization of the GK and AHC genes in the large interval between DMD and DXS28. Cosmid clones and single copy DNA fragments isolated from these two regions are currently being analyzed for expressed sequences by searching for conserved sequences on zoo blots in mouse YAC clones extending from the 3'end of the murine dystrophin gene and by using exon amplification protocols described above (A.P. Walker et al., unpubl.).

GENE LOCI FOR XK, CYBB, RP3, AND OTC IN PROXIMAL Xp21

Clinical phenotypes

CYBB is a severe disease in which patients have recurring bacterial and fungal infections, granulomatous infiltration of multiple organs, and early deaths (for reviews, see Orkin 1989; Smith and Curnutte 1991). In most families, CYBB is inherited as an X-linked recessive trait, but about 30% of cases exhibit an autosomal recessive inheritance pattern. Affected patients and carriers could be identified by a test for decreased activity of the NADPH oxidase system using the reduction of a compound called nitroblue tetrazolium. The gene was first localized to the Xp21 region by DNA linkage analysis in families segregating the CYBB phenotype and by analysis of the contiguous deletion syndrome found in patient NF (Baehner et al. 1986). The X-linked gene for CYBB isolated by Royer-Pokora et al. (1986) is organized in about 25–30 kb of genomic DNA (Orkin 1989) and encodes for the B-chain of cytochrome b_{245}, a component of the NADPH oxidase system (Dinauer et al. 1987; Teahan et al. 1987). Using two polymorphisms identified with the CYBB cDNA, about 50% of families segregating mutations could be counseled and offered prenatal diagnosis (Battat and Francke 1989; Pelham et al. 1990). Better diagnostic tests for the CYBB gene would require more informative polymorphisms or rapid analysis of its exons for point mutations.

McLeod syndrome (XK), named after Hugh McLeod, the original patient in whom it was discovered (Allen et al. 1961), is a red blood cell antigen defect with weak expression of the Kell antigens and a complete absence of the Kx antigen. The lack of the Kx antigen of approximately 37 kD (Redman et al. 1988) gives rise to spiny-shaped erythrocytes called acanthocytes (Marsh 1978; Symmans et al. 1979; Marsh and Redman 1990). Acanthocytes have a decreased life span, resulting in sequestration in the spleen, hemolysis, and splenomegaly. XK patients can also exhibit hepatomegaly and various neuromuscular manifestations (for review, see Witt et al. 1992). They show elevated creatine kinase levels, progressive cardiomyopathy, and a late-onset progressive neuropathy and choreic movements. Muscle biopsy usually shows mild myopathic changes, and dystrophin has been shown to be normal (Carter et al. 1990; Danek et al. 1990).

X-linked retinitis pigmentosa is characterized by night blindness with onset in the first or second decade followed by progressive contraction of the visual fields with loss of central vision before the fourth decade, diffuse retinal pigmentary changes, and absence of electroretinal responses. It was first thought to be a single disorder, since linkage studies originally showed a location on the proximal short arm of the X chromosome (Bhattacharya et al. 1984). It became clear from further linkage studies that there were at least two distinct but clinically identical forms of retinitis pigmentosa on the short arm, termed RP3 and RP2 (Nussbaum et al. 1985; Wright et al. 1987; Denton et al. 1988; Musarella et al. 1988, 1990; Wirth et al. 1988; Chen et al. 1989; Ott et al. 1990). The majority of cases are associated with an RP3 locus distal to OTC, within or near the Xp21 deletion of the patient BB (Francke et al. 1985), whereas a minority of patients show that the RP2 locus is located more proximally between DXS7 and DXS14 (Ott et al. 1990). As mentioned previously, there is also linkage evidence for a third RP locus (RP6) between DMD and DXS28 (Musarella et al. 1988, 1990).

OTC deficiency is a severe urea cycle disorder resulting in hyperammonemia, low plasma citrulline, and high levels of orotic acid in the urine, with death occurring soon after birth in severe cases (Brusilow and Horwich 1989). It was first linked to CYBB in a female patient with mild mental retardation and heterozygosity for OTC deficiency and CYBB who exhibited a cytogenetically detectable deletion within Xp21 (Francke 1984). OTC deficiency was also associated with AHC and GK deficiency, since DNA analyzed from a male patient who died 36 hours after birth was found to contain a deletion in Xp21 (Hammond et al. 1985). The cDNA clone isolated for the human OTC gene was found to be 1.6 kb and was predicted to encode a protein product of 354 amino acids (Horwich et al. 1984). The human OTC gene was organized into ten exons spread over 73 kb of genomic DNA (Hata et al. 1988). In a minority of cases, the OTC gene was found to be completely or partially deleted (Rozen et al. 1985) and was used in prenatal diagnosis by direct gene analysis by Old et al. (1985). Using the OTC cDNA, point mutations have been detected in infants with OTC deficiency (for review, see Grompe et al. 1990).

Physical mapping of XK, CYBB, RP3, and OTC

Although the deletion identified in patient BB showed that XK, CYBB, and RP3 were all in Xp21, analysis of deletions found to occur in DNA of other patients gave rise to the gene order in Xp21 as pter-DMD-XK-CYBB-RP3-OTC-cen (Figs. 1 and 5). Molecular analysis of DNA from two patients, SB with XK, CYBB, and RP3 (de Saint-Basile et al. 1988), and OM with XK and CYBB (Frey et al. 1988), showed that these three genes were in close proximity and that if the clinical diagnoses were true, RP3 was not located between XK and CYBB. Additionally, a patient with DMD only and a cytogenetically detectable deletion (JD; Wilcox et al. 1986), and two first cousins (S/H) with XK only, allowed the positioning of XK distal to CYBB (Bertelson et al. 1988). From the pattern of probes deleted, it was estimated that XK, CYBB, and RP3 were in proximal Xp21 relatively far from the 5'end of DMD (Fig. 5).

Deletion junction fragments isolated from the DNA of two patients (BB and JD) with large cytogenetically visible deletions facilitated the construction of a PFGE map of the region containing XK, CYBB, and RP3 (Musarella et al. 1991; Ho et al. 1992). Both patients had one deletion breakpoint in the DMD gene. Therefore, a distal flanking DMD probe could be used in each case to screen a genomic library constructed from patient DNA. Musarella et al. (1991) isolated the BB deletion junction, and a fragment from the proximal side (XH1.4, DXS1082) was used in conjunction with the CYBB cDNA to construct a PFGE map.

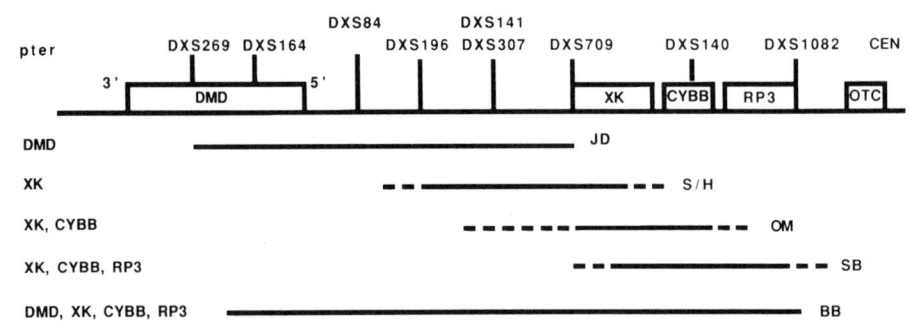

Figure 5 Schematic diagram showing the order of genes, anonymous DNA probes, and patient deletions in proximal Xp21. The phenotypes of the patients are shown on the left side of the deletions; dashed lines at the ends of the S/H, OM, and SB deletions indicate the limits of uncertainty of their breakpoints. DXS1082 (XH1.4) and DXS709 (3BHR0.3) are the proximal deletion junction fragments isolated from patients BB and JD (Ho et al. 1992; Musarella et al. 1991, respectively).

This map defined an approximately 150–170-kb region that should contain the RP3 gene. This area had a CpG island, about 35 kb distal to the BB deletion breakpoint, that is a candidate site for the RP3 gene. Ho et al. (1992) isolated the deletion junction fragment from the patient JD, who exhibited only DMD, to provide a flanking marker for XK, since this patient was missing some of the Xp21 probes also deleted in the McLeod syndrome cousins (S/H; Fig. 5). A junction fragment on the proximal side of the JD deletion breakpoint (DXS709) was used in conjunction with the CYBB cDNA to construct a 1.1-Mbp PFGE map. This map showed a 700-kb region with multiple CpG islands and limited the McLeod syndrome critical region to 150–380 kb distal to CYBB. The region between XK and DMD seems to be quite large (>2 Mbp), yet no distinct phenotype is exhibited when it is deleted. This suggests that there are few genes in this region, or that their functions are redundant in genes elsewhere in the genome.

A YAC contig containing the XK, CYBB, RP3, and OTC genes

Using the JD junction fragment (DXS709), CYBB cDNA, and OTC cDNA as hybridization probes on the ICRF human YAC library, five YAC clones were isolated and characterized, and formed a 1.7-Mbp YAC contig (Fig. 6) (Ho et al. 1991). One YAC clone of 850 kb contained the DXS709 and CYBB loci, and another YAC clone of 650 kb contained the CYBB and first 5 exons of the OTC gene. This mapped CYBB and OTC to within 450 kb and oriented the OTC gene transcription unit on the chromosome as pter-5'-3'-cen. The three YAC clones containing XK, CYBB, RP3, and the 5'end of OTC have been used as hybridization probes to screen the ICRF X chromosome cosmid filters. Positive cosmid

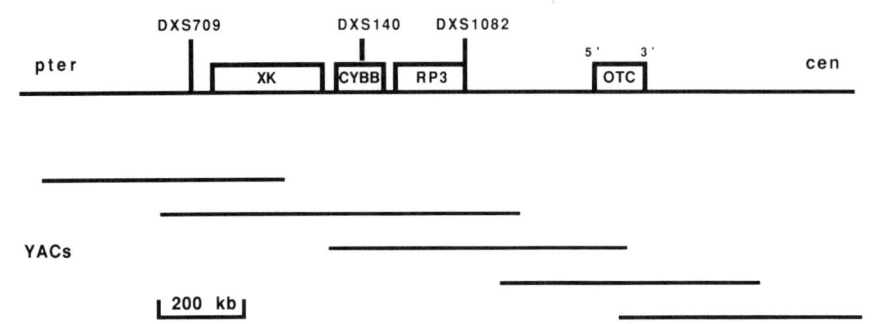

Figure 6 Diagram of a 1.7-Mbp YAC contig containing the genes for XK, CYBB, RP3, and OTC (Ho et al. 1991). The limits of the XK and RP3 critical regions are drawn to scale, but the genomic sizes of the CYBB and OTC genes are not (25–30 kb and 73 kb, respectively). The proximal deletion junction of patient BB (XH1.4, DXS1082; Musarella et al. 1991) is shown as the proximal border of the RP3 locus critical region but has not been hybridized to these YAC clones. The exact position of the ends of the YAC clones relative to each other is still being investigated, and the possibility of chimeric ends has not been determined for all clones (M.F. Ho et al., unpubl.).

clones were identified and placed in physically defined intervals according to the hybridization pattern determined with each YAC. The YAC and cosmid clones from this region are currently being analyzed for expressed sequences that may be candidate genes for XK and RP3 (M.F. Ho et al., unpubl.).

CONCLUSION

The present state of the physical map for the region Xp21 has been defined by analyzing deletion breakpoints in contiguous deletion syndrome patients and by constructing long-range PFGE maps and, more recently, YAC contigs. The current state of the map is such that all disease genes are precisely ordered with respect to each other. The large 2.4-Mbp DMD gene has been isolated in overlapping YAC clones and then reconstructed on one YAC. The genes not yet identified for AHC, GK, XK, and RP3 have been isolated on YAC clones, and by deletion analysis the critical regions can all be narrowed to regions of several hundred kilobases. The search for expressed sequences from these critical regions is an ongoing activity in this region using several different methods.

References

Abbs, S., R.G. Roberts, C.G. Mathew, D.R. Bentley, and M. Bobrow. 1990. Accurate assessment of intragenic recombination frequency within the

Duchenne muscular dystrophy gene. *Genomics* 7: 602.
Alitalo, T., T.A. Kruse, H. Forsius, A.W. Eriksson, and A. de la Chapelle. 1991. Localization of the Åland island eye disease locus to the pericentromeric region of the X chromosome by linkage analysis. *Am. J. Hum. Genet.* 48: 31.
Allen, F.H., S.M.R. Krabbe, and P.A. Corocoran. 1961. A new phenotype (McLeod) in the Kell blood-group system. *Vox. Sang.* 6: 555.
Baehner, R.L., L.M. Kunkel, A.P. Monaco, J.L. Haines, P.M. Conneally, C. Palmer, N. Heerema, and S.H. Orkin. 1986. DNA linkage analysis of X-linked chronic granulomatous disease. *Proc. Natl. Acad. Sci.* 83: 3398.
Ballabio, A. 1991. Contiguous deletion syndromes. *Curr. Opin. Genet. Dev.* 1: 25.
Bar, S., E. Barnea, Z. Levy, S. Neuman, D. Yaffe, and U. Nudel. 1990. A novel product of the Duchenne muscular dystrophy gene which differs from the known isoforms in its structure and tissue distribution. *Biochem. J.* 272: 557.
Bartley, J. and C. Gies. 1989. A Xp21 deletion assigns locus DXS28 (C7) proximal to DXS68 (L1.4) and DXS67 (B24) and has the proximal breakpoint in the intron 3' to the first exon of DMD-8 (DMD exon 47). *Cytogenet. Cell. Genet.* 51: 958.
Bartley, J.A., P. Shivanand, S. Davenport, D. Goldstein, and J. Pickens. 1986. Duchenne muscular dystrophy, glycerol kinase deficiency, and adrenal insufficiency associated with Xp21 interstitial deletion. *J. Pediatr.* 108: 189.
Battat, L. and U. Francke. 1989. NsiI RFLP at the X-linked chronic granulomatous disease locus (CYBB). *Nucleic Acids Res.* 17: 3619.
Baumbach, L.L., J.S. Chamberlain, P.A. Ward, N.J. Farwell, and C.T. Caskey. 1989. Molecular and clinical correlations of deletions leading to Duchenne and Becker muscular dystrophies. *Neurology* 39: 465.
Beggs, A.H., M. Koenig, F.M. Boyce, and L.M. Kunkel. 1990. Detection of 98% of DMD/BMD deletions by PCR. *Hum. Genet.* 86: 45.
Bertelson, C.J., A.O. Pogo, A. Chaudhuri, W.L. Marsh, C.M. Redman, D. Banerjee, W.A. Symmans, T. Simon, D. Frey, and L.M. Kunkel. 1988. Localization of the McLeod locus (XK) within Xp21 by deletion analysis. *Am. J. Hum. Genet.* 42: 703.
Bhattacharya, S.S., A.L. Wright, J.F. Clayton, W.H. Price, C.L. Phillips, C.M. McKeown, J.M. Bird, P.L. Pearson, E.M. Southern, and H.J. Evans. 1984. Close genetic linkage between X-linked retinitis pigmentosa and a recombinant DNA probe L1.28. *Nature* 309: 253.
Bies, R.D., S.F. Phelps, M.D. Cortez, R. Roberts, C.T. Caskey, and J.S. Chamberlain. 1992. Human and murine dystrophin mRNA transcripts are differentially expressed during skeletal muscle, heart and brain development. *Nucleic Acids Res.* 20: 1725.
Blake, D.L., D.R. Love, J. Tinsley, G.E. Morris, H. Turley, K. Gatter, G. Dickson, Y.H. Edwards, and K.E. Davies. 1992. Characterization of a 4.8kb transcript from the Duchenne muscular dystrophy locus expressed in Schwannoma cells. *Hum. Mol. Genet.* 1: 103.
Blonden, L.A.J., J.T. Den Dunnen, H.M.B. Van Paassen, M.C. Wapenaar, P.M. Grootscholten, H.B. Ginjaar, E. Bakker, P.L. Pearson, and G.J.B. Van Ommen. 1989. High resolution deletion breakpoint mapping in the DMD-gene by whole cosmid hybridization. *Nucleic Acids Res.* 17: 5611.
Boyce, F.M., A.H. Beggs, C. Feener, and L.M. Kunkel. 1991. Dystrophin is tran-

scribed in brain from a distant upstream promoter. *Proc. Natl. Acad. Sci.* **88:** 1276.

Boyd, Y., E. Munro, P. Ray, R. Worton, A.P. Monaco, L.M. Kunkel, and I. Craig. 1987. Molecular heterogeneity of translocations associated with muscular dystrophy. *Clin. Genet.* **31:** 265.

Brusilow, S., and A.L. Horwich. 1989. Urea cycle enzymes. In *The metabolic basis of inherited disease*, 6th ed. (ed. C.R. Scriver et al.), p. 629. McGraw-Hill, New York.

Buckler, A.J., D.D. Chang, S.L. Graw, J.D. Brook, D.A. Haber, P.A. Sharp, and D.E. Housman. 1991. Exon amplification: A strategy to isolate mammalian genes based on RNA splicing. *Proc. Natl. Acad. Sci.* **88:** 4005.

Burghes, A.H.M., C. Logan, X. Hu, B. Belfall, R. Worton, and P.N. Ray. 1987. Isolation of a cDNA clone from the region of an X;21 translocation that breaks within the Duchenne/Becker muscular dystrophy gene. *Nature* **328:** 434.

Burke, D.T., G.F. Carle, and M.V. Olson. 1987. Cloning of large DNA segments of exogenous DNA into yeast by means of artificial chromosome vectors. *Science* **236:** 806.

Burmeister, M. and H. Lehrach. 1986. Long-range restriction map around the Duchenne muscular dystrophy gene. *Nature* **324:** 582.

Burmeister, M., A.P. Monaco, E.F. Gillard, G.-J.B. Van Ommen, N.A. Affara, M.A. Ferguson-Smith, L.M. Kunkel, and H. Lehrach. 1988. A 10 megabase map of human Xp21 including the Duchenne muscular dystrophy gene. *Genomics* **2:** 189.

Carter, N.D., J.E. Morgan, A.P. Monaco, M.S. Schwartz, and S. Jeffery. 1990. Dystrophin expression and genotypic analysis of two cases of benign X linked myopathy (McLeod's syndrome). *J. Med. Genet.* **27:** 345.

Chamberlain, J.S., R.A. Gibbs, J.E. Ranier, P.N. Nguyen, and C.T. Caskey. 1988. Deletion screening of the Duchenne muscular dystrophy locus via multiplex DNA amplification. *Nucleic Acids Res.* **16:** 11141.

———. 1990. Multiplex PCR for the diagnosis of Duchenne muscular dystrophy. In *PCR protocols* (ed. M.A. Innes et al.), p. 272. Academic Press, New York.

Chelly, J., J-P. Concordet, J-C. Kaplan, and A. Kahn. 1989. Illegitimate transcription: transcription of any gene in any cell type. *Proc. Natl. Acad. Sci.* **86:** 2617.

Chelly, J., F. Marlhens, B. Dutrillaux, G.J. Van Ommen, M. Lambert, B. Haioun, G. Boissinot, and M. Fardeau. 1988 Deletion proximal to DXS68 locus (L1 probe site) in a boy with Duchenne muscular dystrophy, glycerol kinase deficiency, and adrenal hypoplasia. *Hum. Genet.* **78:** 222.

Chelly, J., H. Gilgenkrantz, M. Lambert, G. Hamard, P. Chafey, D. Recan, P. Katz, A. de la Chapelle, M. Koenig, I.B. Ginjaar, M. Fardeau, F. Tome, A. Kahn, and J.-C. Kaplan. 1990. Effect of dystrophin gene deletions on mRNA levels and processing in Duchenne and Becker muscular dystrophies. *Cell* **63:** 1239.

Chen, J.-D., F. Halliday, G. Keith, L. Sheffield, P. Dickinson, and P.L. Pearson. 1989. Linkage heterogeneity between X-linked retinitis pigmentosa and a map of 10 RFLP loci. *Am. J. Hum. Genet.* **45:** 401.

Coffey, A.J., R.G. Roberts, E.D. Green, C.G. Cole, R. Butler, R. Anand, F. Giannelli, and D.R. Bentley. 1992. Construction of a 2.6 Mb contig in yeast artificial chromosomes spanning the human dystrophin gene using an

STS-based approach. *Genomics* **12**: 474.
Collins, F.S. 1992. Positional cloning: Let's not call it reverse anymore. *Nature Genet.* **1**: 3.
Cotton, R.G.H., N.R. Rodrigues, and R.D. Campbell. 1988. Reactivity of cytosine and thymidine in single-base-pair mismatches with hydroxylamine and osmium tetroxide and its application to the study of mutations. *Proc. Natl. Acad. Sci.* **85**: 4397.
Coulson, A., R. Waterston, J. Kiff, J. Sulston, and Y. Kohara. 1988. Genome linking with yeast artificial chromosomes. *Nature* **335**: 184.
Danek, A., T.N. Witt, H.B.A.C Stockmann, B.J. Weiss, D.L. Schotland, and K.H. Fischbeck. 1990. Normal dystrophin in Mcleod myopathy. *Annu. Neurol.* **28**: 720.
Darras, B.T. and U. Francke. 1988. Myopathy in complex glycerol kinase deficiency patients is due to 3' deletions of the dystrophin gene. *Am. J. Hum. Genet.* **43**: 126.
Darras, B.T., P. Blattner, J.F. Harper, A.J. Spiro, and U. Francke. 1988. Intrageneic deletions in 21 Duchenne muscular dystrophy (DMD)/Becker muscular dystrophy (BMD) families studied with the dystrophin cDNA: Location of breakpoints on HindIII and BglII exon-containing fragment maps, meiotic and mitotic origin of the mutations. *Am. J. Hum. Genet.* **43**: 620.
Davies, K.E., J.-L. Mandel, A.P. Monaco, R.L. Nussbaum, and H.F. Willard. 1991. Report of the committee on the constitution of the X chromosome. *Cytogenet. Cell. Genet.* **58**: 853.
Davies, K.E., M.N. Patterson, S.J. Kenwrick, M.V. Bell, H.R. Sloan, J.A. Westman, L.J. Elsas II, and J. Mahan. 1988. Fine mapping of glycerol kinase deficiency and congenital adrenal hypoplasia within Xp21 on the short arm of the human X chromosome. *Am. J. Med. Genet.* **29**: 557.
Den Dunnen, J.T., E. Bakker, E.G. Klein-Breteler, P.L. Pearson, and G.J.B. Van Ommen. 1987. Direct detection of more than 50% Duchenne muscular dystrophy mutations by field inversion gel electrophoresis. *Nature* **329**: 640.
Den Dunnen, J.T., P.M. Grootscholten, E. Bakker, L.A.J. Blonden, H.B. Ginjaar, M.C. Wapenaar, H.M.B. van Paassan, C. van Broeckhoven, P.L. Pearson, and G.J.B. van Ommen. 1989. Topography of the Duchenne muscular dystrophy gene: FIGE and cDNA analysis of 194 cases reveals 115 deletions and 13 duplications. *Am. J. Hum. Genet.* **45**: 835.
Den Dunnen, J.T., P.M. Grootscholten, J.G. Dauwerse, A.P. Walker, A.P. Monaco, R. Butler, R. Anand, A.J. Coffey, D.R. Bentley, H.Y Steensma, and G.J.B. Van Ommen. 1992. Reconstruction of the 2.4 Mb human DMD-gene by homologous YAC reconstruction. *Hum. Mol. Genet.* **1**: 19.
Denton, M.J., J.-D. Chen, S. Serravalle, P. Colley, F.B., Halliday, and J. Donald. 1988. Analysis of linkage relationships of X-linked retinitis pigmentosa with the following Xp loci: L1.28, OTC, 754, XJ1.1, pERT87, and C7. *Hum. Genet.* **78**: 6.
de Saint-Basile, G., M.C. Bohler, A. Fischer, J. Cartron, J.L. Dufier, C. Griscelli, and S.H. Orkin. 1988. Xp21 DNA microdeletion in a patient with chronic granulomatous disease, retinitis pigmentosa and McLeod phenotype. *Hum. Genet.* **80**: 85.
Dinauer, M., S.H. Orkin, R. Brown, A.J. Jesaitis, and C.A Parkos. 1987. The

glycoprotein encoded by the X-linked chromic granulomatous disease locus is a component of the neutrophil cytochrome b complex. *Nature* **327**: 717.

Dunger, D.B., M. Pembrey, P. Pearson, A. Whitfield, K.E. Davies, B. Lake, D. Williams, and M.J.D. Dillon. 1986. Deletion on the X chromosome detected by direct DNA analysis in one of two unrelated boys with gylcerol kinase deficiency, adrenal hypoplasia, and Duchenne muscular dystrophy. *Lancet* **I**: 585.

Duyk, G.M., S. Kim, R.M. Myers, and D.R. Cox. 1990. Exon trapping: A genetic screen to identify candidate transcribed sequences in cloned mammalian genomic DNA. *Proc. Natl. Acad. Sci.* **87**: 8995.

Elvin, P., G. Slynn, D. Black, A. Graham, R. Butler, J. Riley, R. Anand, and A.F. Markham. 1990. Isolation of cDNA clones using yeast artificial chromosome probes. *Nucleic Acids Res.* **18**: 3913.

Ervasti, J.M. and K.P. Campbell. 1991. Membrane organization of the dystrophin-glycoprotein complex. *Cell* **66**: 1.

Feener, C.A., M. Koenig, and L.M. Kunkel. 1989. Alternative splicing of human dystrophin mRNA generates isoforms at the carboxy terminus. *Nature* **338**: 509.

Forrest, S.M., G.S. Cross, A. Speer, D. Gardner-Medwin, J. Burn, and K. Davies. 1987. Preferential deletion of exons in Duchenne and Becker muscular dystrophies. *Nature* **329**: 638.

Francke, U. 1984. Random X inactivation resulting in mosaic nullisomy of region Xp21.1-21.3 associated with heterozygosity for ornithine transcarbamylase deficiency and for chronic granulomatous disease. *Cytogenet. Cell Genet.* **38**: 298.

Francke, U., J.F. Harper, B.T. Darras, J.M. Cowan, E.R.B. McCabe, A. Kohlschutter, W.K. Seltzer, F. Saito, J. Goto, J.P. Harpey, and J.E. Wise. 1987. Congenital adrenal hypoplasia, myopathy, and glycerol kinase deficiency: Molecular genetic evidence for deletions. *Am. J. Hum. Genet.* **40**: 212.

Francke, U., H.D. Ochs, B. de Martinville, J. Giacalone, V. Lindgren, C.M. Dieteche, R.A. Pagon, M.H. Hofker, G.J.B. Van Ommen, P.L. Pearson, and R.J. Wedgwood. 1985. Minor Xp21 chromosome deletion in a male associated with expression of Duchenne muscular dystrophy, chronic granulomatous disease, retinitis pigmentosa and the McLeod syndrome. *Am. J. Hum. Genet.* **37**: 250.

Frey, D., M. Machler, R. Seger, W. Schmid, and S. Orkin. 1988. Gene deletion in a patient with chronic granulomatous disease and McLeod syndrome: Fine mapping of the Xk gene locus. *Blood* **71**: 252.

Frohman, M.A., M.K. Dush, and G.R. Martin. 1988. Rapid amplification of full-length cDNAs from rare transcripts: Amplification using a single gene-specific oligonucleotide primer. *Proc. Natl. Acad. Sci.* **85**: 8998.

Gòrecki, D.C., A.P. Monaco, J.M.J. Derry, A.P. Walker, E.A. Barnard, and P.J. Barnard. 1992. Expression of four alternative dystrophin transcripts in brain regions regulated by different promoters. *Hum. Mol. Genet.* **1**: 505.

Grompe, M., S.N. Jones, and C.T. Caskey. 1990. Molecular detection and correction of ornithine transcarbamylase deficiency. *Trends Genet.* **6**: 335.

Hammond, J., N.J. Howard, R. Brookwell, S. Purvis-Smith, B. Wilcken, and N. Hoogenraad. 1985. Proposed assignment of loci for X-linked adrenal hypoplasia and glycerol kinase genes. *Lancet* **I**: 54.

Hata, A., T. Tsuzuki, K. Shimada, M. Takiguchi, M. Mori, and I. Matsuda. 1988. Structure of the human ornithine transcarbamylase gene. *J. Biochem.* **103**: 302.

Heilig, R., C. Lemaire, and J.-L. Mandel. 1987. A 230 kb cosmid walk in the Duchenne muscular dystrophy gene: Detection of a conserved sequence and of a possible deletion prone region. *Nucleic Acids Res.* **15**: 9129.

Herrmann, B., S. Labeit, A. Poutska, T.R. King, and H. Lehrach. 1990. Cloning of the T gene required in mesoderm formation in the mouse. *Nature* **343**: 617.

Ho, M.F., A.P. Monaco, L.A.J. Blonden, G.J.B. Van Ommen, N.A. Affara, M.A. Ferguson-Smith, and H. Lehrach. 1992. Fine mapping of the McLeod locus (XK) to a 150-380 kb region in Xp21. *Am. J. Hum. Genet.* **50**: 317.

Ho, M.F., G.A.P. Bruns, N.A. Affara, M.A. Ferguson-Smith, L.A.J. Blonden, G.J.B. Van Ommen, H. Lehrach, and A.P. Monaco. 1991. Physical mapping of the McLeod locus and isolation of a 1.7 Mb YAC contig containing the genes for McLeod, chronic granulomatous disease (CGD), retinitis pigmentosa form 3 (RP3) and ornithine transcarbamylase (OTC). *Cytogenet. Cell. Genet.* **58**: A27094.

Horwich, A.L., W.A. Fenton, K.R. Williams, F. Kalousek, J.P. Kraus, R.F. Doolittle, W. Koningsberg, and L.E. Rosenberg. 1984. Structure and expression of a complementary DNA for the nuclear coded precursor of human mitochondrial ornithine transcarbamylase. *Science* **224**: 1068.

Hu, X, P.N. Ray, E.G. Murphy, M.W. Thompson, and R.G. Worton. 1990. Duplicational mutation at the Duchenne muscular dystrophy locus: Its frequency, distribution, origin, and phenotype genotype correlation. *Am. J. Hum. Genet.* **46**: 682.

Hugnot, J.P., H. Gilgenkrantz, N. Vincent, P. Chafey, G.E. Morris, A.P. Monaco, Y. Berwald-Netter, A. Koulakoff, J.-C. Kaplan, A. Kahn, and J. Chelly. 1992. Novel products of the dystrophin gene: A distal transcript initiated from a unique alternative first exon encoding a 75 kDa protein widely distributed in non-muscle tissues. *Proc. Natl. Acad. Sci.* **89**: 7506.

Huxley, C. and A. Gnirke. 1991. Transfer of yeast artificial chromosomes from yeast to mammalian cells. *BioEssays* **13**: 545.

Ibraghimov-Beskrovnaya, O., J.E. Ervasti, C.J. Leveille, C.A. Slaughter, S.W. Sernett, and K.P. Campbell. 1992. Primary structure of dystrophin-associated glycoproteins linking dystrophin to the extracellular matrix. *Nature* **355**: 696.

Kenwrick, S., M. Patterson, A. Speer, K. Fischbeck, and K.E. Davies. 1987. Molecular analysis of the Duchenne muscular dystrophy region using pulsed-field gel electrophoresis. *Cell* **48**: 351.

Koenig, M. and L.M. Kunkel. 1990. Detailed analysis of the repeat domain of dystrophin reveals four potential hinge segments that may confer flexibility. *J. Biol. Chem.* **265**: 4560.

Koenig, M., A.P. Monaco, and L.M. Kunkel. 1988. The complete sequence of dystrophin predicts a rod-shaped cytoskeletal protein. *Cell* **53**: 219.

Koenig, M., E.P. Hoffman, C.J. Bertelson, A.P. Monaco, C. Feener, and L.M. Kunkel. 1987. Complete cloning of the Duchenne muscular dystrophy (DMD) cDNA and preliminary genomic organization of the DMD gene in normal and affected individuals. *Cell* **50**: 509.

Koenig, M., A.H. Beggs, M. Moyer, S. Scherpf, K. Heindrich, T. Bettecken, G.

Meng, C.R. Muller, M. Lindlof, H. Kaariainen, A. de la Chapelle, A. Kiuru, M.-L. Savontaus, H. Gilgenkrantz, D. Recan, J. Chelly, J.-C. Kaplan, A.E. Covone, N. Archidiacono, G. Romeo, S. Liechti-Gallati, V. Scheider, S. Braga, H. Moser, B.T. Darras, P. Murphy, U. Francke, J.D. Chen, G. Morgan, M. Denton, C.R. Greenberg, K. Wrogemann, L.A.J. Blonden, H.M.B. van Passan, G.J.B. van Ommen, and L.M. Kunkel. 1989. The molecular basis for Duchenne versus Becker muscular dystrophy: Correlation of severity with type of deletion. *Am. J. Hum. Genet.* **45**: 498.

Korn, B., Z. Sedlacek, A. Manca, P. Kioschis, D. Konecki, H. Lehrach, and A. Poustka. 1992. A strategy for the selection of transcribed sequences in the Xq28 region. *Hum. Mol. Genet.* **1**: 235.

Koussef, B. 1981. Linkage between chronic granulomatous disease and Duchenne muscular dystrophy? *Am. J. Dis. Child.* **135**: 1149.

Kunkel, L.M., A.P. Monaco, W. Middlesworth, H. Ochs, and S.A. Latt. 1985. Specific cloning of DNA fragments absent from the DNA of a male patient with an X-chromosome deletion. *Proc. Natl. Acad. Sci.* **82**: 4778.

Kunkel, L.M. and 72 coauthors. 1986. Analysis of deletions in DNA from patients with Becker and Duchenne muscular dystrophy. *Nature* **322**: 73.

Larin, Z., A.P. Monaco, and H. Lehrach. 1991. Yeast artificial chromosome libraries containing large inserts from mouse and human DNA. *Proc. Natl. Acad. Sci.* **88**: 4123.

Lederfein, D., Z. Levy, N. Augier, D. Mornet, G. Morris, O. Fuchs, D. Yaffe, and U. Nudel. 1992. A 71-kilodalton protein is a major product of the duchenne muscular-dystrophy gene in brain and other nonmuscle tissues. *Proc. Natl. Acad. Sci.* **89**: 5346.

Legouis, R., J.-P. Hardelin, J. Levilliers, J-M. Claverie, S. Compain, V. Wunderle, P. Millasseau, D. LePaslier, D. Cohen, D. Caterina, L. Bougueleret, H. Delemarre-Van deWaal, G. Lutfalla, J. Weissenbach, and C. Petit. 1991. The candidate gene for X-linked Kallmann syndrome encodes a protein related to adhesion molecules. *Cell* **67**: 423.

Lindsay, S. and A.P. Bird. 1987. Use of restriction enzymes to detect potential gene sequences in mammalian DNA. *Nature* **327**: 336.

Loh, E.Y., J.F. Elliott, S. Cwirla, L.L. Lanier, and M.M. Davis. 1989. Polymerase chain reaction with single-sided specificity: Analysis of T cell receptor δ chain. *Science* **243**: 217.

Love, D.R., J.F. Bloomfield, S.J. Kenwrick, J.R.W. Yates, and K.E. Davies. 1990. Physical mapping distal to the DMD locus. *Genomics* **8**: 106.

Lovett, M., J. Kere, and L.M. Hinton. 1991. Direct selection: A method for the isolation of cDNAs encoded by large genomic regions. *Proc. Natl. Acad. Sci.* **88**: 9628.

McCabe, E.R.B. 1989. Disorders of glycerol metabolism. In *The metabolic basis of inherited disease*, 6th ed. (ed. S.C Scriver et al.), p. 945. McGraw-Hill, New York.

McCabe, E.R.B., J. Towbin, J. Chamberlain, L. Baumbach, J. Witkowski, G.J.B. vanOmmen, M. Koenig, L.M. Kunkel, and W.K. Seltzer. 1989. Complementary DNA probes for the Duchenne muscular dystrophy locus demonstrate a previously undetectable deletion in a patient with dystrophic myopathy, glycerol kinase deficiency, and congenital adrenal hypoplasia. *J. Clin. Invest.* **83**: 95.

Malhotra, S.B., K.A. Hart, H.J. Klamut, N.S.T. Thomas, S.E. Bodrug, A.H.M.

Burghes, M. Bobrow, P.S. Harper, M.W. Thompson, P.N. Ray, and R.G. Worton. 1988. Frame-shift deletions in patients with Duchenne and Becker muscular dystrophy. *Science* **242**: 756.

Marsh, W.L., 1978. Chronic granulomatous disease, the McLeod syndrome, and the Kell blood groups. *Birth Defects* **14**: 9.

Marsh, W.L. and W.C. Redman. 1990. The Kell blood group system: A review. *Transfusion* **30**: 158.

Meitinger, T., Y. Boyd, R. Anand, and I.W. Craig. 1988. Mapping of Xp21 translocation breakpoints in and around the DMD gene by pulsed field gel electrophoresis. *Genomics* **3**: 315.

Monaco, A.P. 1989. Dystrophin, the protein product of the Duchenne/Becker muscular dystrophy gene. *Trends Biochem. Sci.* **14**: 412.

Monaco, A.P., C.J. Bertelson, C. Colletti-Feener, and L.M. Kunkel. 1987. Localization and cloning of Xp21 deletion breakpoints involved in muscular dystrophy. *Hum. Genet.* **75**: 221.

Monaco, A.P., C.J. Bertelson, S. Liechti-Gallati, H. Moser, and L.M. Kunkel. 1988 An explanation for the phenotypic difference between patients bearing partial deletions of the DMD locus. *Genomics* **2**: 90.

Monaco, A.P., A.P. Walker, I. Millwood, Z. Larin, and H. Lehrach. 1992. A yeast artificial chromosome contig containing the complete Duchenne muscular dystrophy gene. *Genomics* **12**: 465.

Monaco, A.P., R.L. Neve, C. Colletti-Feener, C.J. Bertelson, D.M. Kurnit, and L.M. Kunkel. 1986. Isolation of candidate cDNAs for portions of the Duchenne muscular dystrophy gene. *Nature* **323**: 646.

Monaco, A.P., C.J. Bertelson, W. Middlesworth, C.-A. Colletti, J. Aldridge, K.H. Fischbeck, R. Bartlett, M.A. Pericak-Vance, A.D. Roses, and L.M. Kunkel. 1985. Detection of deletions spanning the Duchenne muscular dystrophy locus using a tightly linked DNA segment. *Nature* **316**: 842.

Moser, H. 1984. Duchenne muscular dystrophy: Pathogenetic aspects and genetic prevention. *Hum. Genet.* **66**: 17.

Musarella, M.A., L. Anson-Cartwright, S.M. Leal, L.D. Gilbert, R.G. Worton, G.A. Fishman, and J. Ott. 1990. Multipoint linkage analysis and heterogeneity testing in twenty X-linked retinitis pigmentosa families. *Genomics* **8**: 286.

Musarella, M.A., C.L. Anson-Cartwright, C. McDowell, A.H.M. Burghes, S.E. Coulson, R.G. Worton, and J.M. Rommens. 1991. Physical mapping at a potential X-linked retinitis pigmentosa locus (RP3) by pulsed-field gel electrophoresis. *Genomics* **11**: 263.

Musarella, M.A., A. Burghes, L. Anson-Cartwright, M.M. Mahtani, R. Argonza, L.-C. Tsui, and R. Worton. 1988. Localization of the gene for X-linked recessive type of retinitis pigmentosa (XLRP) to Xp21 by linkage analysis. *Am. J. Hum. Genet.* **43**: 484.

Nizetic, D., G. Zehetner, A.P. Monaco, L. Gellen, B.D. Young, and H. Lehrach. 1991. Construction, arraying and high density display of large insert libraries of the human chromosomes X and 21: Their potential use as reference libraries. *Proc. Natl. Acad. Sci.* **88**: 3233.

Nussbaum, R.L., R.A. Lewis, J.G. Lesko, and R. Ferrell. 1985. Mapping of ophthalmological disease. II. Linkage of relationships of X-linked retinitis pigmentosa to X chromosome short arm markers. *Hum. Genet.* **70**: 45.

Old, J.M., S. Purvis-Smith, B. Wilcken, P. Pearson, R. Williamson, P.L. Briand, N.J. Howard, J. Hammond, L. Cathelineau, and K.E. Davies. 1985. Prena-

tal exclusion of ornithine transcarbamylase deficiency by direct gene analysis. *Lancet* I: 73.
Olson, M.V., L. Hood, C.R. Cantor, and D. Botstein. 1989. A common language for physical mapping of the human genome. *Science* **245:** 1434.
Orkin, S. 1989. Molecular genetics of chronic granulomatous disease. *Annu. Rev. Immunol* **7:** 277.
Ott, J., S. Bhattacharya, J.D. Chen, M.J. Denton, J. Donald, C. Dubay, G.J. Farrar, G.A. Fishman, D. Frey, A. Gal, P. Humphries, B. Jay, M. Jay, M. Litt, M. Machler, M. Musarella, M. Neugebauer, R.L. Nussbaum, J.D. Terwilliger, R.G. Weleber, B. Wirth, F. Wong, R.G. Worton, and A.F. Wright. 1990. Localizing multiple X chromosome-linked retinitis pigmentosa loci using multilocus homogeneity tests. *Proc. Natl. Acad. Sci.* **87:** 701.
Patil, S.R., J.A. Bartley, J.C. Murray, V.V. Ionasescu, and P.L. Pearson. 1985. X-linked glycerol kinase, adrenal hypoplasia and myopathy maps at Xp21. *Cytogenet. Cell Genet.* **40:** 720.
Parimoo, S., S.R. Patanjali, H. Shukla, D.D. Chaplin, and S.M. Weissman. 1991. cDNA selection: Efficient PCR approach for the selection of cDNAs encoded in large chromosomal DNA fragments. *Proc. Natl. Acad. Sci.* **88:** 9623.
Pelham, A., M.-A.J. O'Reilly, S. Malcolm, R.J. Levinsky, and C. Kinnon. 1990. RFLP and deletion analysis for X-linked chronic granulomatous disease using the cDNA probe: Potential for improved prenatal diagnosis and carrier determination. *Blood* **76:** 820.
Pillers, D.-A.M., J.A. Towbin, J.S. Chamberlain, D. Wu, J. Ranier, B.R. Powell, and E.R.B. McCabe. 1990. Deletion mapping of Aland Island eye disease to Xp21 between DXS67 (B24) and Duchenne muscular dystrophy. *Am. J. Hum. Genet.* **47:** 795.
Rapaport, D., D. Lederfein, J.T. Den Dunnen, P.M. Grootscholten, G.J.B. Van Ommen, O. Fuchs, U. Nudel, and D. Yaffe. 1992. Characterization and cell type distribution of a novel, major transcript of the Duchenne muscular dystrophy gene. *Differentiation* **49:** 187.
Ray, P.N., B. Belfall, C. Duff, C. Logan, V. Kean, M.W. Thompson, J.E. Sylvester, J.L. Gorski, R.D. Schmickel, and R.G. Worton. 1985. Cloning of the breakpoint of an X;21 translocation associated with Duchenne muscular dystrophy. *Nature* **318:** 672.
Redman, C.M., W.L. Marsh, A. Scarborough, C.L. Johnson, B.I. Rabin, and M. Overbeeke. 1988. Biochemical studies on McLeod phenotype red cells and isolation of Kx antigens. *Br. J. Haemotol.* **68:** 131.
Roberts, R.G., M. Bobrow, and D.R. Bentley. 1992a. Point mutations in the dystrophin gene. *Proc. Natl. Acad. Sci.* **89:** 2331.
Roberts, R.G., A.J. Coffey, M. Bobrow, and D.R. Bentley. 1992b. Determination of the exon structure of the distal portion of the dystrophin gene by vectorette PCR. *Genomics* **13:** 942.
Roberts, R.G., T.F.M. Barby, E. Manners, M. Bobrow, and D.R. Bentley. 1991. Direct detection of dystrophin gene rearrangements by analysis of dystrophin mRNA in peripheral blood lymphocytes. *Am. J. Hum. Genet.* **49:** 298.
Royer-Pokora, B., L.M. Kunkel, A.P. Monaco, S.C. Goff, P.E. Newburger, R.L. Baehner, F.S. Cole, J.T. Curnette, and S.H. Orkin. 1986. Cloning the gene for an inherited human disorder—chronic granulomatous disease—on the

basis of its chromosomal location. *Nature* **322**: 32.
Rozen, R., J. Fox, W.A. Fenton, A.L. Horwich, and L.E. Rosenberg. 1985. Gene deletion and restriction fragment length polymorphisms at the human transcarbamylase locus. *Nature* **313**: 815.
Schmickel, R.D. 1986. Contiguous gene syndromes: A component of recognizable syndromes. *J. Pediatr.* **109**: 231.
Smith, R.M. and J.T. Curnette. 1991. Molecular basis of chronic granulomatous disease. *Blood* **77**: 673.
Symmans, W.A., C.S. Shepherd, W.L. Marsh, R. Oyen, S.B. Shohet, and B.J. Linehan. 1979. Hereditary acanthocytosis associated with the McLeod phenotype of the Kell blood group system. *Br. J. Haematol.* **42**: 575.
Teahan, C., P. Rowe, P. Parker, N. Totty, and A.W. Segal. 1987. The X-linked chronic granulomatous disease gene codes for the Beta-chain of the cytochrome b-245. *Nature* **327**: 720.
Towbin, J.A., J.S. Chamberlain, D. Wu, D.-A.M. Pillers, W.K. Seltzer, and E.R.B. McCabe. 1990. DXS28(C7) maps centromeric to DXS68(L1-4) and DXS67(B24) by deletion analysis. *Genomics* **7**: 442.
Trask, B.J., H.F. Massa, and M. Burmeister. 1992. Fluorescence in situ hybridization establishes the order cen-DXS28 (C7)-DXS67(B24)-DXS68(L1)-tel in human chromosome Xp21.3. *Genomics* **13**: 455.
Uberbacher, E.E. and R.J. Mural. 1991 Locating protein-coding regions in human DNA sequences by a multiple sensor-neural network approach. *Proc. Natl. Acad. Sci.* **88**: 11262.
Van Ommen, G.J.B., J.M.H. Verkerk, M.H. Hofker, A.P. Monaco, L.M. Kunkel, P. Ray, R. Worton, B. Wieringa, E. Bakker, and P.L. Pearson. 1986. A physical map of 4 million base pairs around the Duchenne muscular dystrophy gene on the human X-chromosome. *Cell* **47**: 499.
Van Ommen, G.J.B., C.E. Bertelson, H.B. Ginjaar, J.T. Den Dunnen, E. Bakker, J. Chelly, M. Matton, A.J. Van Essen, J. Bartley, L.M. Kunkel, and P.L. Pearson. 1987. Long-range genomic map of the Duchenne muscular dystrophy (DMD) gene: Isolation and use of J66 (DXS268), a distal intragenic marker. *Genomics* **1**: 329.
Walker, A.P., Z. Larin, H. Lehrach, and A.P. Monaco. 1991. Human Xp21 YAC contig maps. *Cytogenet. Cell Genet.* **58**: 2087.
Walker, A.P., J. Chelly, D.R. Love, Y. Ishikawa Brush, D. Récan, J-L. Chaussain, C.A. Oley, J.M. Conner, J. Yates, D.A. Price, M. Super, A. Bottani, B. Steinmann, J-C. Kaplan, K.E. Davies, and A.P. Monaco. 1992. A YAC contig in Xp21 containing the adrenal hypoplasia congenita and glycerol kinase deficiency genes. *Hum. Mol. Genet.* **1**: 579.
Wapenaar, M.C., T. Kievits, K.A. Hart, S. Abbs, L.A.J. Blonden, J.T. Den Dunnen, P.M. Grootscholten, E. Bakker, C. Verellen-Dumoulin, M. Bobrow, G.J.B. Van Ommen, and P.L. Pearson. 1988. A deletion hot spot in the Duchenne muscular dystrophy gene. *Genomics* **2**: 101.
Weleber, R.G., D.M. Pillers, C.E. Hanna, B. Powell, R.E. Magenis, and N.R.M. Buist. 1989. Åland Island disease (Forsius-Eriksson syndrome) associated with contiguous gene syndrome at Xp21: Similarity to incomplete stationary night blindness. *Arch. Ophthalmol.* **107**: 1170.
Wieringa, B., T. Hustinx, J. Scheres, W. Ranier, and B. ter Haar. 1985. Complex glycerol kinase deficiency syndrome explained as X-chromosomal deletion. *Clin. Genet.* **27**: 522.

Wilcox, D.E., A. Cooke, J. Colgan, E. Boyd, A. Aitken, L. Sinclair, L. Glasgow, J.B.P. Stephenson, and M.A. Ferguson-Smith. 1986. Duchenne muscular dystrophy due to familial Xp21 deletion detectable by DNA analysis and flow cytometry. *Hum. Genet.* **73**: 175.

Wirth, B., M.J. Denton, J.-D. Chen, M. Neugebauer, F.B. Halliday, M. van Schooneveld, J. Donald, E.M. Bleeker-Wagemakers, P.L. Pearson, and A. Gal. 1988. Two different genes for X-linked retinitis pigmentosa. *Genomics* **2**: 263.

Witt, T.N., A Danek, M. Reiter, M.U. Heim, J. Dirschinger, and E.G.J. Olsen. 1992. McLeod syndrome: A distinct form of neuroacanthocytosis. *J. Neurol.* **239**: 302.

Worley, K.C., J.A. Towbin, X.M. Zhu, D.F. Barker, A. Ballabio, J. Chamberlain, L.G. Biesecker, S.L. Blethen, P. Brosnan, J.E. Fox, W.B. Rizzo, G. Romeo, N. Sakuragawa, W.K. Seltzer, S. Yamaguchi, and E.R.B. McCabe. 1991. Identification of new markers in Xp21 between DX528(C7) and DMD. *Genomics* **13**: 957.

Wright, A.F., S.S. Bhattacharya, J.F. Clayton, M. Dempster, P. Tippett, C.M.E. McKeown, M. Jay, B. Jay, and A.C Bird. 1987. Linkage relationships between X-linked retinitis pigmentosa and nine short-arm markers: Exclusion of the disease locus from Xp21 and localization to between DXS7 and DXS14. *Am. J. Hum. Genet.* **41**: 635.

Yates, J.R.W., E.F. Gillard, A. Cooke, J.M. Colgan, T.J. Evans, and M.A. Ferguson-Smith. 1987. A deletion of Xp21 maps congenital adrenal hypoplasia distal to glycerol kinase deficiency. *Cytogenet. Cell Genet.* **46**: 723.

Large-scale DNA Sequence Analyses of Mammalian T-Cell Receptor Loci

Leroy Hood,[1] Ben F. Koop,[2] Lee Rowen,[1] and Kai Wang[1]

[1]Department of Molecular Biotechnology
University of Washington
Seattle, Washington 98105

[2]Department of Biology
Center for Environmental Health
University of Victoria
Victoria, British Columbia V8W 2Y2

Large-scale DNA sequence analysis provides a powerful approach for the analysis of chromosomal information. We discuss the justifications for genomic sequence analysis as well as its application to the T-cell receptor loci of humans and mice.

Topics discussed include:

- ❏ justification for large-scale genomic DNA sequencing
- ❏ rationale for selecting regions for large-scale DNA sequencing
- ❏ T-cell receptor biology
- ❏ methodology of large-scale DNA sequencing
- ❏ an analysis of the 100-kb $C_\delta C_\alpha$ region of the mouse α/δ T-cell receptor locus
- ❏ an evolutionary comparison of a position of the $C_\delta C_\alpha$ regions of humans and mice

INTRODUCTION

During the past 4.5 billion years, a remarkable biological information-handling system has evolved in living organisms. The primary repository of information is the DNA molecule. The one-dimensional units of information or genes inscribed in DNA with its 4-letter alphabet are converted into the three-dimensional structures of proteins with their 20-letter alphabet. The expression of genes is controlled by adjacent regulatory sequences such as promoters, enhancers, and silencers. The regulatory regions of genes and many proteins exhibit a remarkable network of interactions to generate the necessary information for development—that fascinating process whereby humans, for example, start as a single cell, the fertilized zygote, and mature to an organism with 10^{14} cells exhibiting a myriad of different phenotypes.

Remarkably, most genes are not contiguous units of information in plants and animals, but rather, most are broken up into coding regions (exons) and noncoding regions (introns), an organization that presumably reflects the primordial evolution of the distinct functional motifs of information assembled into contemporary genes. The genes are transcribed directly into nuclear RNA molecules and these, after splicing out the intervening intron sequences, generate messenger RNA molecules, which, in turn, are translated into protein molecules.

Many of the major functions of DNA are carried out through the one-dimensional molecular complementarity of base-pairing (replication, transcription, propagation of mutations, etc.). In contrast, the order of the 20 amino acid residues of proteins specifies how they fold in three dimensions to constitute molecular machines that catalyze the chemistry of life and give the body shape and form.

The DNA molecule is the backbone of an organelle of information storage, replication, transfer, and evolution, the chromosome. We recognize that the term "organelle" is usually reserved for membrane-bound compartments, but we do believe each chromosome is a "compartment" of information, and hence we have used the term descriptively. The chromosome has a variety of functions, which include the compact folding and unfolding of meters of DNA, movement, replication, and transcription. These functions are mediated by specific DNA/protein interactions that, in part, are reflected, perhaps subtly, in the chromosomal sequences. In a general sense, DNA sequences specify information that falls into three broad categories: coding regions, regulatory elements, and regions mediating the general functions of the chromosome as an informational organelle.

The DNA of the chromosomes, or genomic DNA, has the total gene structure complete with introns and regulatory sequences. In contrast, the transcribed mRNA molecules contain coding regions in addition to varying amounts of 5'- and 3'-flanking sequences; the introns and regulatory regions are not included. In man, it is estimated that there are 50,000–100,000 genes. Accordingly, both genomic sequences and complementary DNA (cDNA) sequences of mRNAs afford important and complementary opportunities to understand the nature of our biological coding system.

The biological information encoded in chromosomes is actively maintained by ongoing evolutionary processes. Hence, the best way to identify chromosomal regions that specify this information is to carry out evolutionary comparisons of species that have diverged sufficiently in their non-informational sequences that conserved informational sequences can be readily recognized. The chromosome carries out many additional functions that have not yet been defined and, at least for some of these functions, the sequence conservation between diverging evolutionary lines could occur over large distances and in rather complex patterns (e.g., the conservation of certain combinations of purine and/or pyrimidine bases). In these cases, evolutionary comparisons, coupled with appropriate computational analyses, may be the key to unraveling these complex patterns, and they may provide subtle clues to the yet undefined functions of these chromosomal DNA sequences.

THE CASE FOR LARGE-SCALE GENOMIC SEQUENCING

Some scientists have argued that a sequence analysis of many cDNA clones from different tissues should be the focus of the large-scale DNA sequencing thrust of the human genome project. On the surface, this suggestion is attractive. Sequencing cDNAs allows one to examine genes directly without analyzing the larger intergenic regions and introns. Moreover, tissue or cellular location of the expressed genes can be defined by appropriate selection of the source tissue for the cDNA library. In this manner, it is suggested that most of the 50,000–100,000 human genes could easily be sequenced in the next few years with several large and concerted efforts.

We believe that cDNA sequencing has several obvious limitations. First, the frequent cDNAs will be sequenced repeatedly unless time-consuming normalization precautions are taken. For example, mRNA molecules may be expressed at levels differing by one million-fold or more. Hence, only the more abundant species will be sequenced in ordinary cDNA libraries. In many libraries, rare genes will never be sequenced. Indeed, this may be the case for more than 50% of the genes. Attempts to equalize the mRNA concentrations, so-called normalization, have met with mixed success. Clearly, genomic DNA provides already normalized libraries with genes present in equimolar mixtures. Second, many genes are expressed in few cells at limited stages of development and, accordingly, may never be included in most cDNA libraries. Hence, there are major concerns about what fraction of the total genes the cDNA approach will actually identify and how efficient the process will be.

We agree that cDNA sequence analyses are one powerful approach to large-scale sequencing of the human genome, but we also think it would be a serious mistake in strategy not to put a substantial effort into the large-scale sequence analysis of genomic DNA. Large-scale genomic DNA sequencing has many potential benefits.

1. We see large-scale genomic DNA sequencing as a multistep process beginning with the physical mapping of the region to be sequenced, the preparation of small DNA fragments for sequencing, the sequence determination of these fragments, and finally, the assembly of the little DNA strings into larger strings of DNA (Fig. 1) (Hunkapiller et al. 1991b). Large-scale genomic sequencing drives the technological development of all steps of this process in a way that the much simpler sequencing of cDNA clones does not. The objective is to improve the throughput of DNA sequencing by 50–100-fold over the next 5 years. If we succeed in this endeavor, sequencing the human genome and many other genomes will become the most efficient way to find all of the species genes. Hence, genomic sequencing is a powerful driver for large-scale DNA sequencing technologies.
2. Genomic sequencing will drive the development of algorithms to identify coding regions and regulatory elements directly from genomic DNA sequence data (cDNA sequences do not contain the regulatory regions). It will also promote the development of algorithms to search for short- and long-range patterns associated with more general chromosomal functions. Driving the general development of informatics is as important as driving sequencing technology development. Large-scale genomic sequencing will also push the development of other requisite computational tools such as signal processing, image analysis, laboratory information management systems, and more effective databases (Hood et al. 1992a).
3. Genomic sequencing is the only way to eventually identify all human genes and, at the same time, the organizational features of all gene families.
4. The human genome project will soon yield a dense 2 centiMorgan (cM) genetic map (1 cM is roughly equivalent to 10^6 base pairs or 1 megabase [Mb]). This genetic map will allow, in principle, the localization of genes to 2-Mb regions. Clearly, the most rapid way to identify the gene of interest in a 2-Mb region is through DNA sequence analysis of this entire region (a 2-cM region may encode 30–60 genes). To highlight this problem, the Huntington's chorea gene has been localized to a 2–4-Mb region at the tip of human chromosome 4 for nine years, yet the gene is still not identified. We believe large-scale

DNA SEQUENCING OF T-CELL RECEPTOR LOCI   **67**

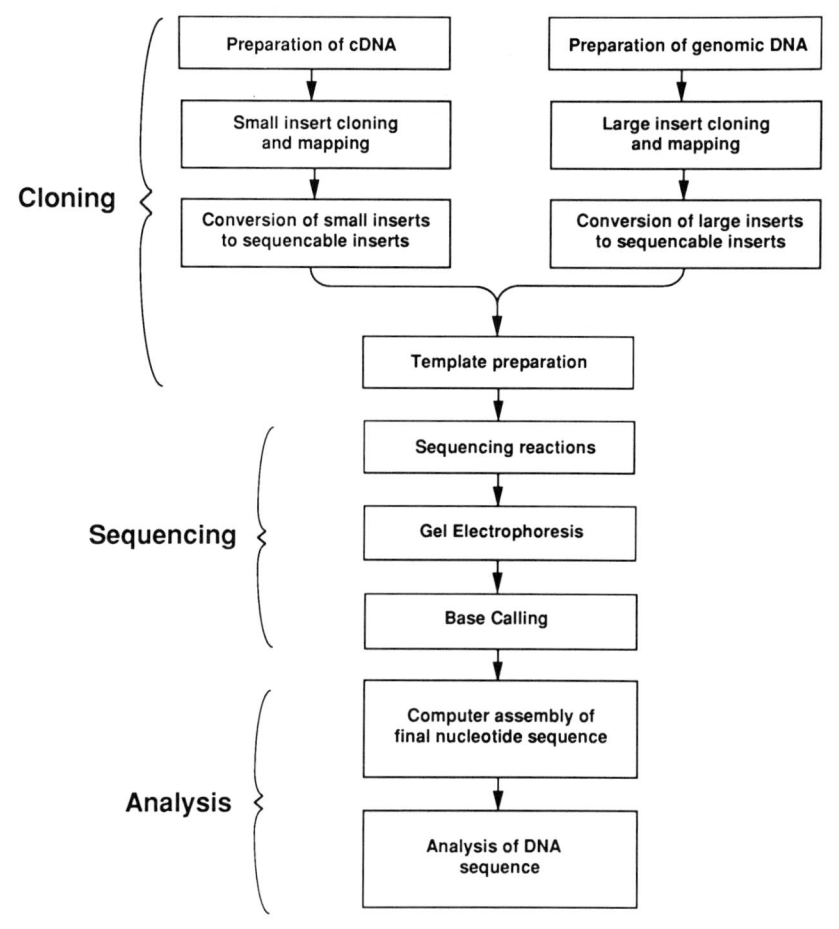

```
Cloning {
    Preparation of cDNA          Preparation of genomic DNA
            ↓                              ↓
    Small insert cloning          Large insert cloning
       and mapping                   and mapping
            ↓                              ↓
  Conversion of small inserts    Conversion of large inserts
   to sequencable inserts         to sequencable inserts
}
                     ↓
            Template preparation
                     ↓
Sequencing {
            Sequencing reactions
                     ↓
            Gel Electrophoresis
                     ↓
              Base Calling
}
                     ↓
Analysis {
         Computer assembly of
         final nucleotide sequence
                     ↓
            Analysis of DNA
               sequence
}
```

Figure 1 Schematic diagram of the steps involved in large-scale DNA sequencing (see text). (Reprinted, with permission, from Hunkapiller et al. 1991a.)

    genomic sequencing of 2-Mb regions will be critical to identifying many interesting human genes localized by genetic mapping.
 5. Large-scale genomic sequencing will permit genetic markers to be placed virtually at will throughout the genome. For example, PCR primers can be chosen for the unique sequences on either side of highly polymorphic simple repeats (e.g., $[CA]_n$). These so-called polymorphic sequence tagged sites (pSTSs) can then be used to join the genetic and sequence maps (Nickerson et al. 1992).
 6. The complete sequence analysis of nonidentical multigene families gives PCR access to any individual member of the mul-

tigene family. With a knowledge of the sequences of all the members of a gene family, PCR primers can be chosen uniquely for the individual members to access their expression as mRNA. This PCR access to expressed mRNAs opens up striking new opportunities to study the parameters of development and environmentally inductive expression for individual members of interesting multigene families. Such specific PCR primers cannot be designed without the knowledge of the sequences of all the family members, because individual primer pairs could amplify two or more similar unsequenced members of the family.

7. A complete sequence of multigene loci will provide knowledge of all restriction enzyme sites. This knowledge can be used to study size polymorphisms and organizational changes within the gene family in different members of the species and will facilitate comparisons of these gene families in closely related species (see below).

8. The complete sequence analysis of multigene families will reveal all the sequences involved in the regulation of the individual gene members. From these, and from the evolutionary comparative analyses of regions likely to contain control elements (based on their conserved sequence patterns), it is possible to begin deciphering the regulatory code of human development (Hood 1992). The regulatory code specifies the temporal and spatial patterns of individual gene expression, as well as the magnitude of the expression. This code includes a knowledge of both the DNA sequences involved and the *trans*-activating protein factors binding to these sequences. Once the regulatory code is broken, biologists will have computer access to these molecular addresses for each gene. This information will provide important clues as to the functions of each gene. Moreover, this regulatory code will allow biologists to identify the networks of coordinately expressed genes that trigger development and the expression of mature cellular phenotypes.

9. A complete identification of the 50,000–100,000 human genes will allow biologists to begin identifying the lexicon of motifs that are the building block components of the three-dimensional structures of proteins (Hood 1992). These fundamental pattern analyses will come not only from comparing all linear protein (gene) sequences, but also from searching with new informatic methods the three-dimensional motifs of solved three-dimensional structures so that the emerging one- and three-dimensional motif patterns can be correlated. This lexicon of one- and three-dimensional motifs will play a valuable role in helping solve the protein folding problem (how a particular

linear sequence of amino acid subunits folds in three dimensions) and will also provide insights into the structures and functions of newly identified genes. In principle, cDNA analyses could also achieve this goal, since they are focused on coding regions. We think they are far less likely to contribute comprehensively to the identification of protein motifs because the cDNA approach will fail to identify many genes.

10. If all the members of a multigene family are characterized, then unique short peptides, synthesized on the basis of the sequence, can be used to generate antibodies specific for the individual members. These antibodies serve as tools to study the sites and developmental stages of specific gene expression and to probe structure-function relationships. Once again, generating member-specific antibodies requires a knowledge of the sequences of all the members of a multigene family.

11. The potential for using genomic sequencing for studying evolution is striking (Hood et al. (1992b). Let us discuss several different scenarios.

    a. A comparison of evolutionary lines that diverged at the mammalian radiation, ~80 million years ago, permits the presumably functionally conserved sequences to be distinguished from the more divergent sequences of most introns and intergenic regions. These conserved regions will aid in identifying coding regions, some regulatory regions, and perhaps other areas associated with the general functions of the chromosome as an informational organelle. Certainly the additional coding, regulatory, and "organelle" sequences will provide valuable insights for the identification of protein motifs and regulatory elements and chromosome-specific sequences. The functional entities may exhibit complex sequence patterns (e.g., those of immunoglobulin homology units; Hunkapiller et al. 1989) that can only be discerned as more comparative data accumulate.

    b. The comparison of human sequences with model organism sequences (e.g., bacteria [5 Mb], yeast [15 Mb], nematode [100 Mb], *Drosophila* [150 Mb]) will identify homologous genes. The functions of these genes can then be studied in these model organisms whose genes, proteins, and even tissues can be experimentally manipulated in vivo more easily and effectively than in humans. This may be a key strategy for understanding the functions of many new sequence-defined human genes.

    c. PCR provides access to interesting regions of related genomes (e.g., humans and higher primates). Hence, the

evolution of genes and gene families over shorter time periods (e.g., 35 million years) can readily be analyzed as to the nature of the dynamic mechanisms that catalyze the fascinating features of molecular evolution, such as coincidental or concerted evolution (Hood et al. 1975).
   d. In very closely related species such as human and chimp (99% similar in most coding sequences), the sequence differences may provide fundamental insights into the nature of the species differences. In this case, it appears likely that the bulk of the evolutionary changes will occur as a consequence of changes in regulatory sequences. It is difficult to imagine how proteins 99% homologous could lead to the profound physical and mental differences seen in these closely related primates, hence, a comparison of the regulatory machinery of brain-specific systems will reveal fascinating insights into the process of speciation.
12. Serendipity will play a fascinating role as the information of the chromosomes is unraveled by sequence analysis. Genes will be found in surprising locations. The organization of some multigene families may contain real surprises. Some of the conserved regions in species comparisons will pose striking puzzles as to their functions.

We have already begun to illustrate the benefits of genomic sequencing with our preliminary large-scale DNA sequencing excursions into one immune receptor locus in mouse and human, as discussed below.

THE CHOICE OF REGIONS FOR LARGE-SCALE DNA SEQUENCING

The human genome contains approximately 3 billion nucleotides. Large-scale DNA sequencing is a complex multistep process that includes preparation of the sequencing templates, sequence determination, and the assembly and interpretation of the resulting information (Fig. 1). The front end of this process, and several aspects of the informatics, currently constitute bottlenecks in this system (Hunkapiller et al. 1991b). Indeed, significant contemporary methods of automated DNA sequencing allow one scientist with a fluorescent DNA sequencer to analyze approximately 18,000 nucleotides per day (Hunkapiller et al. 1991b). The process is still expensive at $1–$4 per base pair (Sulston 1992). Hence, the areas chosen for sequence analysis must be carefully selected for biological, genetic, and evolutionary interest. In this latter regard, we feel that species comparisons are vital to identify regions of conserved sequence and, hence, functional significance. A separate motivation for encouraging large-scale DNA sequencing at this time, as discussed earlier, is to take on sequencing problems that are sufficiently challenging to drive the development of new technologies for systems integration, automation, and optimization of existing approaches to the complex, multistep process of automated sequencing. It should also drive the development of completely new sequencing strategies (Hunkapiller et al. 1991a).

We chose to analyze the three T-cell receptor gene families of mouse and man (Wilson et al. 1992; Koop et al. 1992 and in prep.). T cells play a central role in the immune response because of their ability to recognize foreign macromolecules (antigens) and to facilitate the differentiation of other lymphocytes (Fig. 2). This immune recognition is mediated by T-cell receptors (Hunkapiller et al. 1989; Davis 1990; Jorgensen et al. 1992). These receptors recognize peptide fragments of antigen presented by a second type of immune receptor, those encoded by the major histocompatibility complex (MHC). Hence, the T-cell receptors have the capacity to recognize both peptide antigens and MHC molecules, and this dual recognition is fundamental to the generation of tolerance and the expansion of the immune repertoire itself. There are two categories of T cells defined by distinct heterodimer receptors: The α/β receptor is employed by classic T-cells that recognize most foreign antigens, and the γ/δ receptors are on T cells of uncertain function. The T-cell polypeptides are divided into variable (V)(antigen recognition) and constant(C)(fix to T cells and mediate T-cell activation) regions. The var-

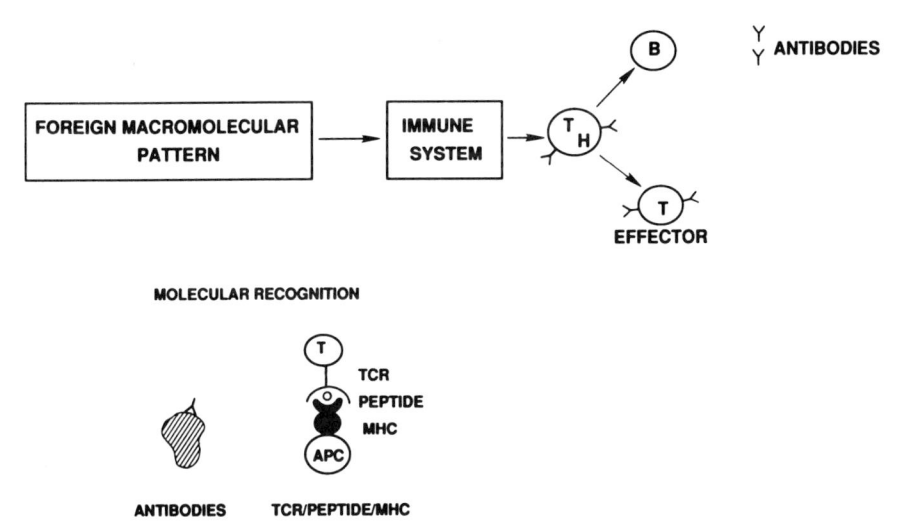

Figure 2 Schematic diagram of the vertebrate immune response to foreign macromolecular patterns (antigens) and the three types of immune receptors, antibodies, and T-cell receptors, and molecules encoded by the MHC. Antibodies generally interact with three-dimensional antigenic structures, and T-cell receptors interact with linear peptide antigens presented by MHC receptors.

iable region is encoded by families of discrete germ-line segments, designated V and joining (J) (α, γ) and V, diversity (D) and J(β, δ). In each T cell there is a rearrangement and contiguous juxtaposition of one set of these gene segments in the germ line to generate VJ or VDJ variable genes. This rearrangement process is mediated by DNA recombination signals adjacent to the gene segments that are joined and by RNA splicing signals that join the 3' end of the J gene segments to the C gene. The C regions are encoded by one or more C genes. The T-cell receptor polypeptides are encoded by three gene families, α/δ, β, and γ, that range in size from 0.2 to 1 Mb (Fig. 3).

Our initial choice for DNA sequence analysis was the ~100-kb region in the α/δ family bounded by the C_δ and C_α genes, respectively (Koop et al. 1992; Wilson et al. 1992). This region is of immunologic interest because the number of J_α gene segments has, up to this point, been unknown. In addition, this region afforded an interesting DNA segment for species comparison (mouse and human) because of its multiplicity of coding elements, DNA rearrangement signals, RNA splicing signals, and two C genes. Moreover, more than 78 different α and δ cDNA sequences, representing 35 different J_α gene segments, are available for comparison against the germ line or chromosomal information.

LARGE-SCALE DNA SEQUENCING STRATEGY

Our present approach to large-scale DNA sequencing has been to employ cosmid inserts as the fundamental unit of DNA sequencing (typically 33–40 kb) and to utilize automated fluorescent DNA sequencing together with the random shotgun approach to sequence analysis (Wilson et al. 1992; B.F. Koop et al., in prep.). We use the ABI 373 fluorescent sequencer, which employs the enzymatic or dideoxy method of DNA se-

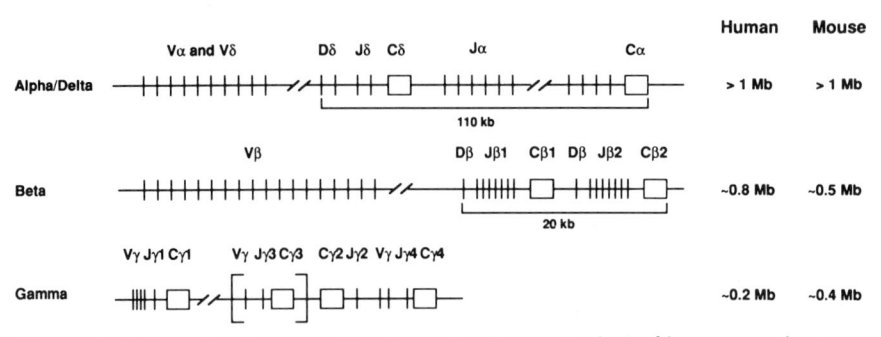

Figure 3 Schematic diagram of the three T-cell-receptor loci of humans and mice. The δ gene family is entirely contained within the α locus. The approximate lengths of these loci are indicated. Vertical lines and open boxes represent gene segment and gene coding regions, respectively.

quencing. Four different fluorescent dyes are coupled to primer sequences or to dideoxynucleotide analogs to color-code the arrays of G, C, A, and T fragments that are generated individually by the enzymatic methodology. We have employed manual methods as well as the Beckman Biomek robotic workstation (Wilson et al. 1988) and the ABI Catalyst (Koop et al. 1993) to carry out the enzymatic sequencing reactions. Since the four fluorescent dyes differentiate sets of fragments terminating in the different bases, the four reaction mixtures can be pooled and run as a single lane on the automated fluorescent sequencer, which directly reads into a computer the color of the bands as they migrate by a certain point on the gel and correspondingly reveals the identity of the nucleotide terminating that particular DNA fragment. Typically, 420-500 base pairs of sequence can be obtained with this methodology (Koop et al. 1993).

There are two general approaches to the sequence analysis of large DNA fragments (Hunkapiller et al. 1991a). The first, directed sequencing, employs the strategy where operationally, one may walk down the cosmid insert sequencing consecutive and overlapping 250-300-base pair stretches of sequence (Siemieniak et al. 1990, 1991). This methodology requires the use of personnel to synthesize DNA primers, carry out timed exonuclease reactions, or randomly place and map transposons inserted in the DNA element. Its advantage is that it does not require extensive computational time to assemble the small DNA sequences into larger ones, and it does not sequence individual regions repeatedly. We have employed the simpler shotgun method of DNA sequencing (Deininger 1983; Bankier and Barrell 1989). We use two different polymerases for DNA sequencing, Taq polymerase and Sequenase (Koop et al. 1993). Each has advantages, but surprisingly, the error rate for both is approximately similar (Fig. 4) (Koop et al. 1993). In this approach, the entire cosmid, with its insert, is randomly sheared into small fragments and size-selected for 0.7-1.5-kb fragments, which are subcloned into bacteriophage M13 vectors. For each cosmid, 700-800 random fragments are sequenced (generating an average eightfold redundant analysis of each base pair), the short DNA fragments are assembled by computational software into increasingly longer DNA strings (contigs), ideally into a single contiguous DNA sequence the size of the cosmid insert, and then sequences are edited to remove as many ambiguities and errors as possible. Generally, we must sequence 1000-1200 fragments to get 800 useful sequences. With eightfold coverage, we sometimes find that a few gaps are left in the sequence. Gaps can be filled by a variety of methods including primer walking and the generation of gap sequence by the polymerase chain reaction if the gap is sufficiently small in size. In Figure 5, we schematically indicate the three cosmids and five plasmid inserts that were sequenced to assemble the 94.6 kb of sequence, which spans the C_δ to C_α regions of the mouse (Koop et

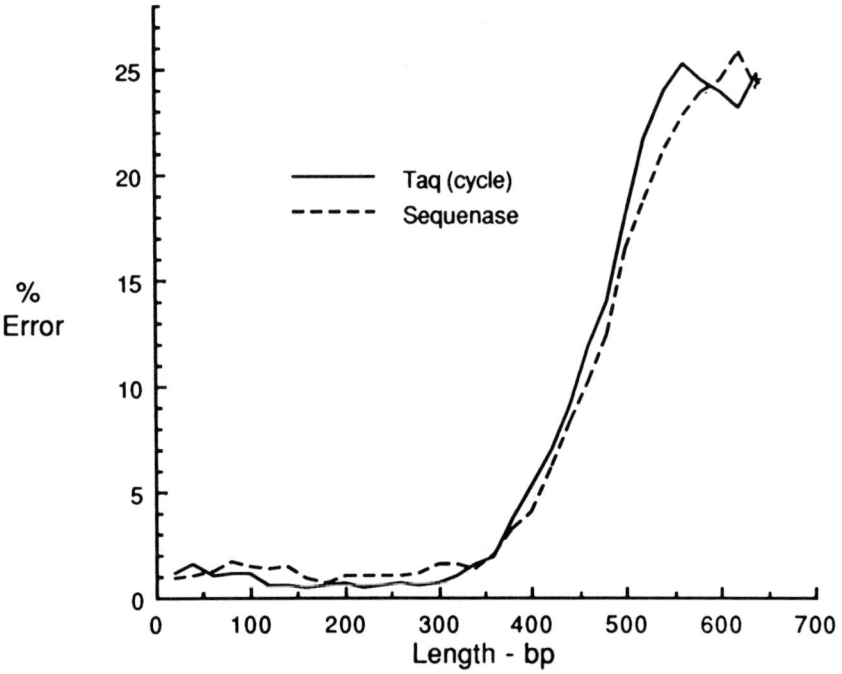

Figure 4 Schematic of the error rates against the length of DNA sequence analyzed for two DNA polymerases routinely employed in enzymatic sequencing reactions. About 400 M13 inserts from one cosmid clone were analyzed for each enzyme. (Reprinted, with permission, from Koop et al. 1993.)

al. 1992; Wilson et al. 1992). This sequence was generated from almost 800 kb of raw DNA sequence data and 120 kb of overlapping finished DNA sequence.

The shotgun method has the virtue of employing simple procedures which are readily automated; it generates great accuracy because of the redundancy; and the operations can be readily mastered by inexperienced personnel. We have determined that our error rate is low, on the order of one in every few thousand nucleotides.

THE $C_\delta C_\alpha$ REGION OF MOUSE

Finding V and J gene segments and C genes

From the literature, more than 78 α and δ cDNA sequences are available (Koop et al. 1992). Previous studies established that the $C_\delta C_\alpha$ region contains the $V_\delta 3$ and at least 35 J_α gene segments, as well as the C_δ and C_α genes. We searched for additional J_α gene segments employing three strategies: (1) a dot-matrix comparison of all human and mouse cDNA

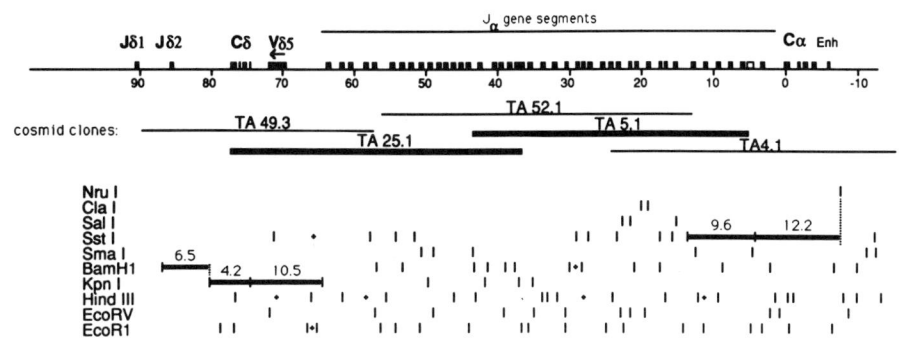

Figure 5 Clone and restriction enzyme map for human $C_\alpha C_\delta$ sequence. The top line represents the consensus physical map of the T-cell receptor $C_\alpha C_\delta$ region with vertical bars indicating coding segments. Thick bars in the two cosmid and five plasmid clones indicate the areas of sequence determination. (Reprinted, with permission, from Wilson et al. 1992.)

J_α sequences against the mouse C_δ C_α region; (2) translation of the mouse C_δ C_α region into protein sequence spanning all three reading frames and then a search for the highly conserved J_α motif Phe-Gly-X-Gly, a sequence present in 90% of the known J_α sequences (Kimura et al. 1987); and (3) a comparison of the 20 kb of then-available human germline region sequence around the C_α gene against its mouse counterpart. We also searched for the 5′ DNA rearrangement sequences (Lewis and Gellert 1981; Hesse et al. 1989), a nonamer (GGTTTTTGT), a 12-base spacer, and a heptamer (CACTGTG), and the consensus 5′ RNA splice site at the 3′ end of the J_α gene segments (CAN^GTAAGT) (Iida 1990). These analyses located the $V_\delta 3$ and the 35 previously characterized J_α gene segments and the C_α and C_δ genes (Fig. 5). They also identified and localized 15 new J_α gene segments, 40 new DNA rearrangement signals, and 40 new RNA splicing signals (Fig. 6).

Are there other genes in the $C_\delta C_\alpha$ region?

There are 69 open reading frames greater than 300 base pair lengths across this stretch of DNA (Koop et al. 1992). Only 3 of these represent open reading frames of identified exons of the C_δ or C_α genes. We can use other criteria to suggest that most of these open reading frames are not functional coding regions of genes (e.g., the lack of appropriate RNA splicing signals). We can also note that the frequency of open reading frames is approximately similar to that of the 73 kb of the human, mouse, and rabbit β-globin loci that have been analyzed to date. A neural net program (CRM/grail) that searches for coding regions based on the asymmetric distribution of 1-, 2-, 3-, and 6-bp sequences identified the first and third exons of the C_δ and the first exon of the C_α gene, but failed to identify the $V_\delta 3$ gene segment (Uberbacker and Mural

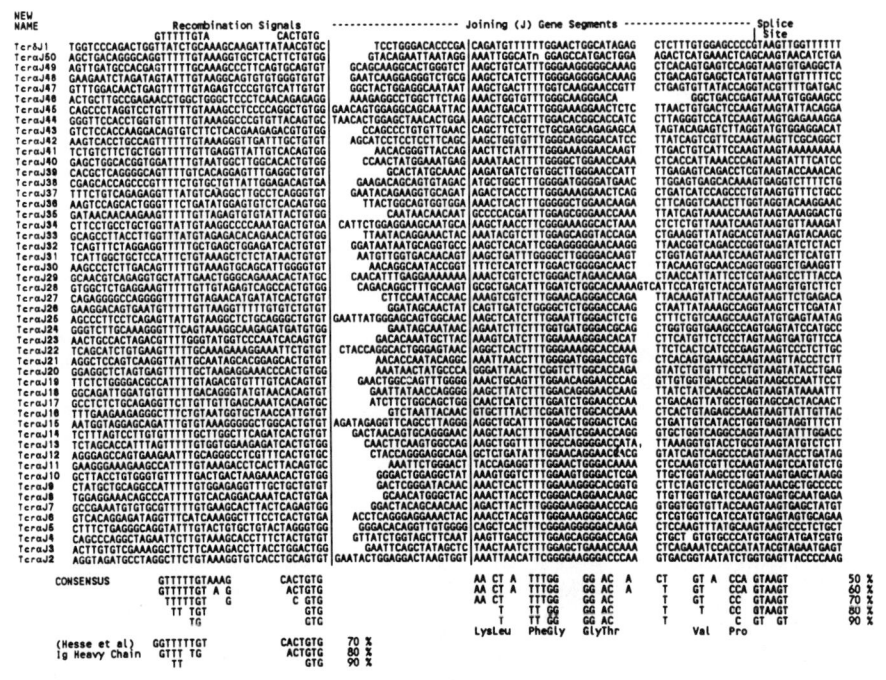

Figure 6 Alignment and consensus sequences for J_α gene regions. The alignment is based on Clustal multiple alignments (Higgins and Sharp 1989) and edited such that it is anchored by the DNA recombination signals (GTTTTTGTA and CACGTGT), J core sequence (TTNGGNNNNGGNAC), and the RNA splice signal (GTAAGT). Sequence divergence in the regions 5' to the DNA recombination signal, 3' to the RNA splice signal, and between the horizontal bars is such that a realistic alignment could not be determined. "n" in TcraJ50 was used for alignment convenience and refers to 11 Ts. (Reprinted, with permission, from Koop et al. 1992.)

1991). This method does not identify coding regions of less than 100 or so bases, hence the J_α gene segments could not be identified. The CRM/grail method identified additional candidate coding regions. Most of these are probably not genes. We expect the computational methods for gene identification to improve significantly in the future. Identifying genes in stretches of genomic sequence poses a serious challenge for which there are currently only unsatisfactory experimental approaches (e.g., PCR analyses of RNA in many tissues, exon trapping). Clearly, the availability of preexisting germ-line and cDNA sequence information was invaluable for the identification of coding regions in the $C_\delta C_\alpha$ region.

Pseudo J_α gene segments

Pseudo J_α segments can generally be identified by five sequence criteria: termination codons within the coding regions, improper reading frames, loss of functional DNA rearrangement sites, loss of functional RNA rearrangement splicing sites, and nonfunctional amino acid substitutions in the coding region. By these criteria, nine J_α gene segments are pseudoelements. In addition, we have assayed the ability of each J_α gene segment to be expressed as a spliced RNA transcript using PCR with a specific PCR primer for each J element and a single C_α primer, thus verifying the functionality of 41 RNA splicing signals. Obviously, the functionality of the DNA rearrangement sequences can be tested in a similar manner. These combined criteria have identified 10 pseudo J elements. There could also be additional pseudo J elements, as relatively few rearranged J gene segments have been verified as functional (e.g., exhibiting antigen-binding functions). In addition, the comparison of the complete mouse and human C_δ C_α region will almost certainly identify additional J_α pseudogenes.

C genes

The striking fact about the C_α and C_δ genes is that they exhibit less than 20% amino acid similarity in their coding regions (Koop et al. 1992). Although their intron-exon organization is similar, these genes have diverged strikingly from one another. This difference is interesting, for it suggests that the C_α and C_δ genes arise from a gene duplication early in the evolution of vertebrate immunity. This divergence poses a fascinating puzzle as to the function of the C_δ gene. As noted above, there are two types of T-cell receptors and T cells, α/β and γ/δ. The α/β T-cell receptors are the classic antigen recognition units. The functions of the γ/δ T-cell receptors remain uncertain.

V gene diversification

V genes are diversified by several distinct mechanisms: (1) diversity of germ-line V, D, and J gene segments; (2) combinatorial joining of the gene elements; (3) sliding boundaries in the sites for joining of the various germ-line elements; and (4) the addition of non-germ-line nucleotides at the junctions between the joined gene segments (this is denoted N [non-germ-line] or P [palindrome] diversity) (Hunkapiller et al. 1989; Davis 1990). The mouse C_δ C_α sequence has shed light on each of these mechanisms: (1) The number of apparently functional murine J_α germ-line gene segments is 40. (2) The number of combinatorial $V_\alpha J_\alpha$ genes is $40n$ where n is the number of functional V_α gene segments. (3 and 4) For the first time, we have defined 40 germ-line J_α gene elements in the mouse and can begin to delineate in detail the nature of the N and P nucleotides added to the gene segment junctions (Koop et al. 1992). An analysis of these data suggests that the sequence of the J_α gene segment influences the nature of bases added by N or P diversification mechanisms (J. Meier, pers. comm.). When the sequences of the complete V_α germ-line repertoire are available, a thorough definition of the N and P nucleotides, as well as the sites of gene segment joining, will provide valuable insights into mechanisms of T-cell receptor diversification.

$V_\delta 3$ gene segment

A single V_δ gene segment is present at the 3' end of the C_δ gene. It is unique in that it is reversed in its transcriptional orientation relative to all the other genes or gene elements defined in this DNA segment. This V_δ gene is functional and must rearrange to its corresponding J_δ gene segment by an inversion event. Whether the unique positioning of this V_δ gene segment has any functional implications is unknown.

The expression of J_α gene segments by PCR analysis

A complete sequence of a large stretch of DNA, together with the delineation of its coding regions, raises the possibility in a multigene family that the transcription of individual coding elements can be examined by PCR (Fig. 7). We have done this in several inbred strains of mice for all of the J_α gene segments present in the T-cell receptor locus (Koop et al. 1992). Nine J_α gene segments are not expressed, and each of these is a pseudogene by two or more features of sequence criteria. In addition, low levels of expression can be delineated; the explanation for these low levels of expression may provide insights to our understanding of immunologic tolerance and selection. The power of large-scale DNA sequencing to characterize the expression of individual members of a multigene family is made possible by the fact that the sequence of each element of the multigene family is known and, accordingly, unique coding sequences can be selected as single-copy DNA primers for PCR. These studies are now being focused on expression of the individual J_α gene elements during differentiation, the nature of the thymic and peripheral T-cell repertoires, and repertoire changes upon immunization and the induction of tolerance. Thus, PCR access to coding regions offers enormous experimental opportunites for studying the molecular biology of multigene families.

Repetitive sequences

Eleven short interspersed nuclear elements (SINEs) have been defined across this 94.6 kb of sequence (Fig. 8) (Lomidze et al. 1986; Deininger 1989; Shakhmuradov and Kolchanov 1989). Ten of these elements are B1, and a single element is a B2 sequence. There is also a single, long interspersed nuclear element (LINE) (Loeb et al. 1986). These repetitive elements have been inserted outside of any coding regions. It is reassur-

DNA SEQUENCING OF T-CELL RECEPTOR LOCI   79

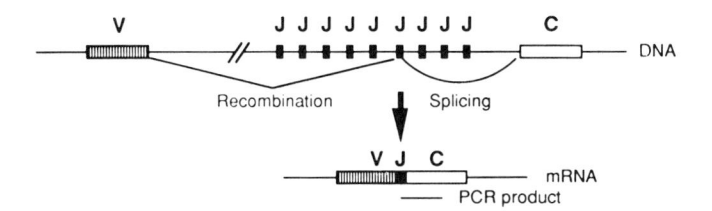

		Mouse strain				Mouse strain	
Tcrα J	B6	DBA2	BALB/C	Tcrα J	B6	DBA2	BALB/C
Ψ1*	-	-	-	26	+	+	+
2	low	low	low	27	+	+	+
Ψ3	-	-	-	28	+	+	+
Ψ4@	+	+	+	Ψ29	-	-	-
5	+	+	+	30	+	+	+
6	+	+	+	31	+	+	+
Ψ7&	-	-	-	32	+	+	+
8	+	+	+	33	+	+	+
Ψ9	-	-	-	34	+	+	+
10	+	+	+	35	+	+	+
11	+	+	+	36	+	+	+
12	+	+	+	37	+	+	+
13	low	low	low	Ψ38	-	-	-
14	+	+	+	Ψ39	-	-	-
15	+	+	+	40	+	+	+
16	+	+	+	41	+	+	+
17	+	+	+	42	+	+	+
18	+	+	+/low	43	+/low	low	+/low
19	+	+	+	44	+	+	+
20#	+	+	+	45	+	+	+
21	+	+	+	Ψ46	-	-	-
22	+	+	+	47	+	+	+
23	+	+	+	48	+	+	+
24	+	+	+	49	+	+	+
25	+	+	+	Ψ50	-	-	-

Figure 7 Transcription of mouse J_α gene segments in thymus and peripheral lymph nodes of C75BL/6, DBA2, and BALB/c inbred mouse strains. An 18- to 20-bp oligonucleotide primer, specific to each of the identified J_α gene segments, was made in addition to a primer from the third exon of the C_α gene. These primers were used to determine whether J_α gene segments were spliced to the C_α gene in the mRNA of the thymus or peripheral lymph nodes in three mouse strains. If the J_α and C_α sequences were spliced and present in mRNA, then a 456-bp PCR product should be observed (+). If the product was detectable but at much reduced level compared to other samples, it was designated as low. All test results were repeated at least twice, and negative and low results were repeated with two independently made primers. It is unclear whether low expression represents a biologically significant phenomenon. In almost all instances, there was little difference between samples taken from the thymus or the peripheral lymph nodes. In the two exceptions (TcrαJ18 and TcrαJ43), thymus results are given first followed by the peripheral node results. (*) Deleted sequences precluded any placement of an oligonucleotide primer. (@) A frameshift mutation does not affect transcription. (&) No feature of the sequence appears to indicate that this J_α gene segment is a pseudogene. (#) A termination codon suggests that this J_α gene segment is a pseudogene, but the codon can be removed by the DNA joining process. (Reprinted, with permission, from Koop et al. 1992.)

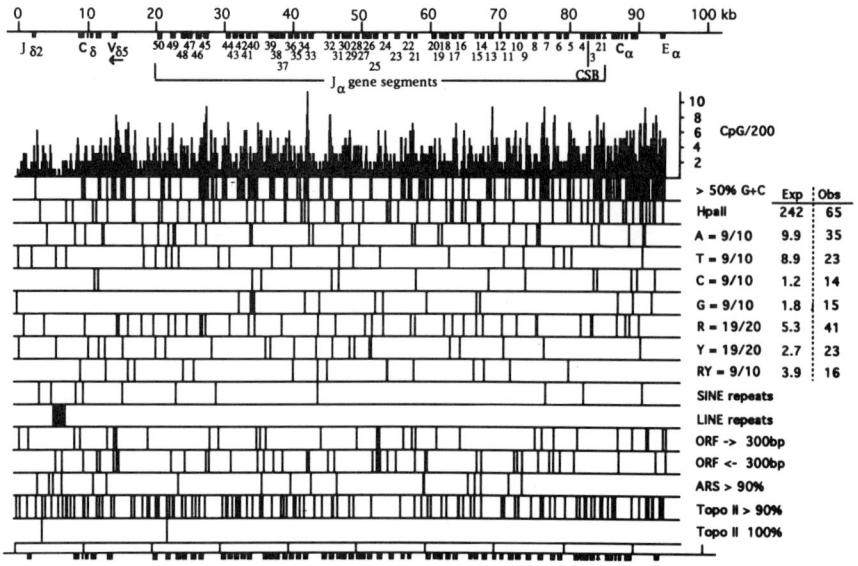

Figure 8 Summary of features found in the mouse C_δ C_α region. The top line is a map of the 95 kb of sequence with all of the known coding regions and other general features (the conserved sequence block [CSB] and α enhancer [E]) marked. The lower lines are positioned with respect to this map. The first line indicates the number of CpG dinucleotides found in a 200-bp window (adjacent windows overlap by 100 bp). The second line marks the position of sequence blocks (200 bp) with greater than 50% G + C content. The third line marks the position of HpaII restriction enzyme sites (CCGG). The distribution of HpaII sites reflects the G + C content as well as the distribution of CpG dinucleotides. The next seven lines mark the position of simple repeats where greater than 9 (19) positions out of 10 (20) are a specified base (R, purines; Y, pyrimidines; RY, alternating purines and pyrimidines). For these features plus HpaII sites, the number of expected occurrences is calculated from the nucleotide frequencies and compared with the observed number of these features. These numbers are shown alongside each of the respective features. The positions of SINEs (B1, B2, and ID repeat sequences) and LINEs are indicated below the simple repeats. The next line indicates the position of open reading frames (ORF) greater than 300 bp in length (all three reading frames are combined, but each strand is presented separately). Finally, the positions of core sequences associated with autonomously replicating sites (ARS) and topoisomerase II sites (TopoII) are indicated. Sites exhibiting greater than 90% or 100% similarity to known core sequences are both positioned with respect to known coding regions (see text). (Reprinted, with permission, from Koop et al. 1992.)

ing to note that the shotgun strategy of DNA sequencing ran into few difficulties in determining the sequences of these repetitive elements. The dispersion of these repetitive elements is typical for their average distribution across the mouse genome. Moreover, in the 20 kb of human C_α sequence available for comparison, the repetitive *Alu* sequences are located in sites distinct from that of the single B1 sequences—consistent with the independent insertion of these elements into the germ line after divergence of the mouse and human evolutionary lines.

DNA SEQUENCING OF T-CELL RECEPTOR LOCI 81

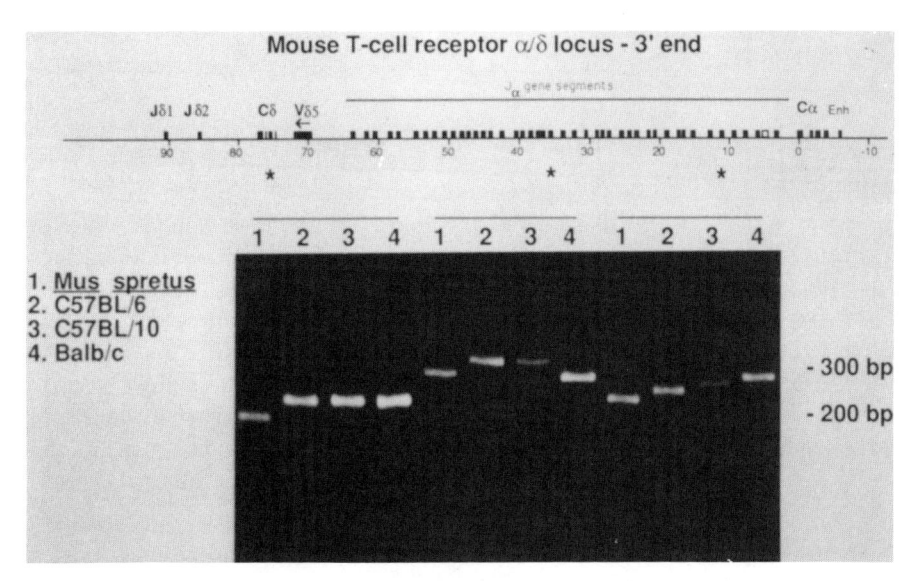

Figure 9 Location and analysis of simple repeat polymorphisms in the mouse C_δ C_α region. The top line indicates the position of coding regions within the 95 kb of mouse sequence. The asterisk marks the location of the repeats tested for polymorphism. The level of polymorphism in each of the three repeats is shown below the physical map. The specific mouse strains used are indicated by the numbers above the gel, and the size of each of the bands can be determined from size markers positioned alongside the gel. (Reprinted, with permission, from Koop et al. 1992.)

Creation of genetic markers

The complete sequence of the C_δ C_α region gives access to the location of simple di-, tri-, and tetranucleotide repeats (Jeffreys et al. 1985; Nakamura et al. 1987; Rogaev 1990). Since these repeats tend to be highly polymorphic in different members of the same species, genetic markers can be created virtually at will by placing unique PCR primers on either side of appropriately selected simple repeats. As a test case, we have placed PCR primers on either side of three repeats in the C_δ C_α region. The sizes of the amplified elements in several inbred strains of mice and in a wild mouse strain are shown in Figure 9. Thus, genetic markers of appropriate location and heterozygosity apparently can be created at will for genetic recombinational analyses.

Use of restriction enzyme sites

The complete DNA sequence of this 94.6-kb stretch of T-cell-receptor locus permits us to choose appropriate restriction sites to analyze the length of this region in closely related mouse species. These studies will be useful both in the evolutionary analysis of closely related species and in the characterization of J_α repertoire polymorphisms within a population of individuals in a given species. In addition, large or small insertions and deletions can readily be characterized across the C_δ C_α region.

Characterization of unusual nucleotide sequences

The C_δ C_α region can be characterized with regard to the distributions of nucleotides present in this sequence (Fig. 8). Are there runs of any one of the four bases or purines or pyrimidines? Are there unusual sequences with particular distribution (ARS, Topo II, matrix attachment sites, etc.)? Are there unusually dense concentrations of the dinucleotide CpG (Bird 1986; Gardiner-Garden and Frommer 1987), indicating the possible 5' boundary of genes? The C_δ C_α region does not have any strikingly unusual features in these regards. However, as we come to understand the sequences necessary for chromosomes to execute their organelle functions, perhaps these analyses will become more revealing.

Evolutionary comparison of the mouse and human C_δ C_α regions

From the literature and from our preliminary sequence analysis of the human C_δ C_α region, we have compared about 20 kb of sequence around the human C_α gene with its mouse counterpart. This region includes a comparison of the C_α gene as well as seven J_α gene segments (Fig. 10). There are six striking observations contained in these data. First, the overall similarity of this region is surprisingly high at 66%. This is surprising because coding regions represent only about 8% of this sequence, hence the noncoding regions are also highly conserved. We must ask ourselves why this noncoding region is conserved. Second, some coding regions are highly conserved (e.g., exons 1 and 3 of the C_α gene) whereas others are not. It is fascinating to consider the forces that may lead to the differential evolution of individual coding regions in members of a multigene family, especially in view of the critical role this gene family plays in vertebrate immunity. Third, the 3'α enhancer element is also highly conserved (Winoto and Baltimore 1989a; Leiden 1991). This observation is important because it suggests that some regulatory elements may be tentatively identified by evolutionary comparisons (this is how the first-identified mammalian enhancer element, the κ immunoglobulin enhancer, was initially identified). However, several other regulatory elements identified 3' to the mouse C_α gene (silencer I and silencer II) do not show any evolutionary conservation (Winoto and Baltimore 1989b). Since the window of nucleotides analyzed is 20, one could argue that perhaps only subelements of this 20-nucleotide window need be conserved, and accordingly, one would not see the conservation. Indeed, perhaps these elements are not conserved in their overall nucleotide sequence but are conserved with regard to subtler features that govern their interactions with regulatory proteins. Fourth, there are short elements of highly conserved DNA sequence scattered across the C_δ C_α region that do not correspond to coding regions. We have studied one of these elements, the conserved sequence block (CSB) (Fig. 10). This element is 95% conserved over 150 base pairs. When this element is used as a probe in Southern blot analyses of vertebrate DNA, it appears highly conserved (Koop et al. 1992). Accordingly, this element presumably has some function. It is not a coding region, as there are termination codons in every reading frame. Perhaps it is a regulatory element, because preliminary experiments suggest this region can bind lymphocyte nuclear proteins. An alternative possibility is that this is a conserved element that relates to some aspect of general chromosomal function. Clearly, one of the challenging aspects of large-scale DNA sequencing will be to delineate functional significance, if any, of highly conserved chromosomal regions. Fifth, there are six regions where gaps must be inserted either in the mouse or human sequences because of repeated sequence that has no homology counterpart in the other species. Each of these regions represents the insertion of a repetitive element, B1 in the

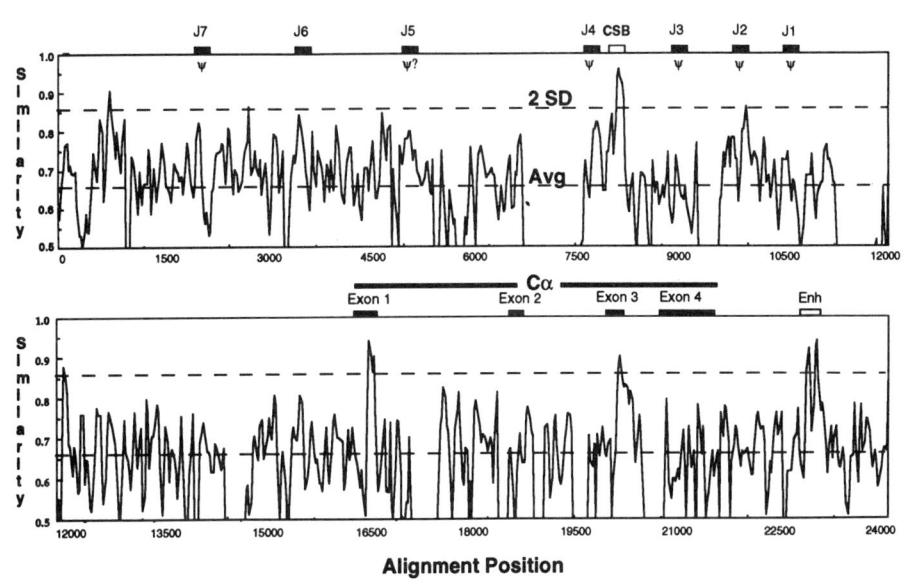

Figure 10 Local similarity plot of the human and mouse C_α regions sequences. Local similarity levels were calculated from a 40-base window that shifted along the alignment at 20-base intervals. In determining levels of similarity, matches were counted as +1, mismatches as 0, and gaps, irrespective of their length, were counted as 0 over 1 base position. With few exceptions, values exceeded 50%. In the case of large gaps, no values were determined. The solid horizontal line indicates the average between the two sequences, and the dashed line indicates similarity values between two standard errors greater than the mean. Similarity plotted in this manner enables regions of high conservation to be identified and subsequently investigated (see text). (Adapted, with permission, from Koop et al. 1992.)

case of mouse and *Alu* in the case of human. Clearly, these elements have been inserted into these evolutionary lines subsequent to their divergence.

Finally, this species comparison led to the identification of a pseudo $J_\alpha 3$ gene segment in the mouse evolutionary line that had not been previously recognized. Accordingly, when the complete human $C_\delta\, C_\alpha$ sequence is compared against its mouse counterpart, there will almost certainly be additional mouse pseudo J_α gene segments identified. Thus, the total size of any multigene family (pseudo and functional genes) can only be determined after careful comparative analyses of evolutionary lines showing appropriate evolutionary divergence.

SUMMARY

Our primary excursion into large-scale DNA sequencing has revealed the power of this new approach toward the delineation of the functional significance of various sequence features of a multigene family. Clearly, the capacity to generate at will PCR primers to interrogate units of RNA expression with regard to biological function (development, repertoire, immunization, tolerance) is perhaps the most striking benefit that will come from the point of view of the functional analysis of immune receptor families. The use of sequences present in the DNA database as well as the general motifs that have been defined in the functional protein and DNA regions will constitute powerful tools for defining functional ele-

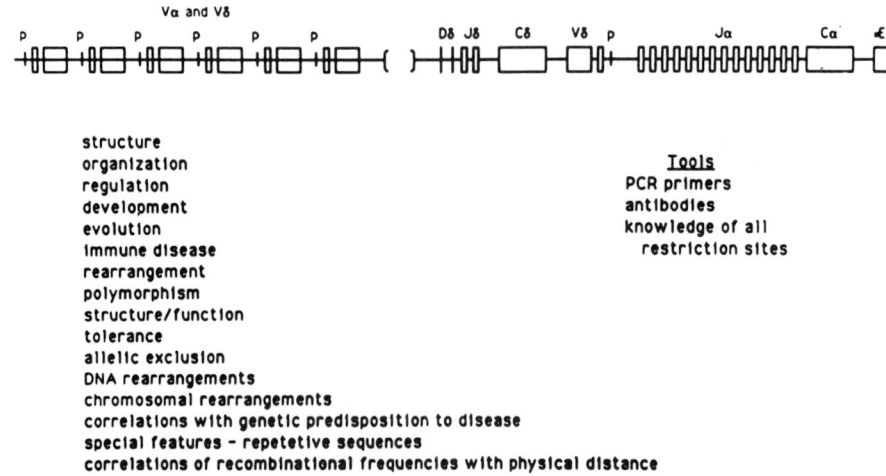

structure
organization
regulation
development
evolution
immune disease
rearrangement
polymorphism
structure/function
tolerance
allelic exclusion
DNA rearrangements
chromosomal rearrangements
correlations with genetic predisposition to disease
special features - repetetive sequences
correlations of recombinational frequencies with physical distance

Tools
PCR primers
antibodies
knowledge of all
restriction sites

Figure 11 Schematic diagram of the human β T-cell-receptor locus. Also included is a list of areas potentially affected by the complete DNA sequence of this region, and a partial list of the "tools" that will arise from the complex sequence analysis of a complex genetic locus.

ments. The ability to create genetic markers with the use of simple repeats will facilitate some fascinating genetic studies (the analysis of recombinational hot spots). The ability to define entirely restriction enzyme sites across stretches of sequences will be useful in comparisons of evolutionary change and polymorphic variability within the species. These analyses will lead to a detailed understanding of the organization of multigene families within a single species. Finally, the possibility of comparative analysis between two or more species leads to observations about the organizations of multigene families (and hence, insights into their evolution) and the identification of conserved regions that may have major functional implications for the delineation of all coding and regulatory regions of the chromosome as an organelle. Indeed, it is important to point out that the comparison of species will allow one to readily identify most of the orthologous sequence present in multigene families through a comparative analysis. It appears that the complete sequence analysis of important complex loci such as the T-cell receptor loci may revolutionize analysis of these loci at virtually every level: structural, organizational, regulatory, developmental, functional, etc. (Fig. 11).

Clearly, the future of large-scale DNA sequencing technology resides in developing an integrated and an automated system with humans largely removed from the operational pathway. When a 100-fold or greater increase in DNA sequencing throughput is achieved, the complete sequence analysis of the human and other genomes will unfold at a pace that is unimaginable today, providing biologists and physicians with an increasing armamentarium of diverse and powerful biological tools.

Acknowledgments

This work was supported by the Department of Energy and the National Institutes of Health. We acknowledge the invaluable assistance of Kim Moulton in typing this manuscript.

References

Bankier, A.T. and B.G. Barrell. 1989. In *Nucleic acids sequencing: A practical approach* (ed. C.M. Howe and C.J. Rawlings), p. 37. IRL Press, Oxford.

Bird, A.P. 1986. CpG-rich islands and the function of DNA methylation. *Nature* **321**: 209.

Davis, M.M. 1990. T-cell receptor gene diversity and selection. *Annu. Rev. Biochem.* **59**: 475.

Deininger, P.L. 1983. Random subcloning of sonicated DNA: Application to shotgun DNA sequence analysis. *Anal. Biochem.* **129**: 216.

———. 1989. SINEs: Short, interspersed repeated DNA elements in higher eukaryotes. In *Mobile DNA* (ed. D.E. Berg and M.M. Howe), p. 619. American Society for Microbiology, Washington D.C.

Gardiner-Garden, M. and M. Frommer. 1987. CpG islands in vertebrate genomes. *J. Mol. Biol.* **196:** 261.

Hesse, J., E. Lieber, K. Mizuuchi, and M. Gellert. 1989. V(D)J recombination—A functional definition of the joining signals. *Genes Dev.* **3:** 1953.

Higgins, D.G. and P.M. Sharp. 1989. Fast and sensitive multiple sequence alignment on a microcomputer. *Gene* **73:** 237.

Hood, L. 1992. Biology and medicine in the twenty-first century. In *The code of codes*, Harvard University Press, Cambridge, Massachusetts.

Hood, L., S. Elgin, and J. Campbell. 1975. The organization, expression and evolution of antibody genes and other multigene families. *Annu. Rev. Genet.* **9:** 305.

Hood, L., J. Solomon, and T. Hunkapiller. 1992a. Computational problems and the human genome project. In *Proceedings of the Fifth SIAM Conference on Parallel Processing for Scientific Computing* (ed. J. Dongarra et al.). Houston, Texas. (In press.)

Hood, L., B.F. Koop, J. Goverman, and T. Hunkapiller. 1992b. Model genomes: The benefits of analyzing homologous human and mouse sequences. *Trends Biol.* **10:** 19.

Hunkapiller, T., J. Goverman, B.F. Koop, and L. Hood. 1989. Molecular evolution in the immunoglobulin gene superfamily. *Cold Spring Harbor Symp. Quant. Biol.* **54:** 15.

Hunkapiller, T., R. Kaiser, B.F. Koop, and L. Hood. 1991a. Large-scale and automated DNA sequence determination. *Science* **254:** 59.

———. 1991b. Large-scale DNA sequencing. *Curr. Opin. Biotechnol.* **2:** 92.

Iida, Y. 1990. Quantification analysis of 5′-splice signal sequences in mRNA precursors. Mutations in 5′-splice signal sequence of human β-globin gene and β thalassemia. *J. Theoret. Biol.* **145:** 523.

Jorgensen, J.L., P.A. Reay, E.W. Ehrich, and M.M. Davis. 1992. Molecular components of T-cell recognition. *Annu. Rev. Immunol.* **10:** 835.

Kimura, N., B. Toyonaga, Y. Yoshikai, R. Du, and T. Mak. 1987. Sequences and repertoire of the human T-cell receptor α and β chain variable region genes in thymocytes. *Eur. J. Immunol.* **17:** 375.

Koop, B.F., L. Rowen, H. Lee, P. Deshpande, and L. Hood. 1993. Advantages and disadvantages of cycle and standard sequencing methods: A statistical analysis. *BioTechniques* (in press).

Koop, B.F., R.K. Wilson, K. Wang, B. Vernooij, D. Zaller, C.L. Kuo, D. Seto, M. Toda, and L. Hood. 1992. Organization, structure and function of 95 kb spanning the murine T-cell receptor Cα to Cδ region. *Genomics* **13:** 1209.

Leiden, J.M. 1991. Transcriptional regulation during T-cell development: The alpha TCR gene as a molecular model. *Immunol. Today* **13:** 22.

Lewis, S. and M. Gellert. 1989. The mechanism of antigen receptor gene assembly. *Cell* **59:** 585.

Loeb, D.D., R.W. Padgett, S.C. Hardies, W.R. Shehee, M.B. Comer, M.H. Degeli, and C.A. Hutchinson III. 1986. The sequence of large L1md element reveals a tandem repeated 5′end and several features found in retrotransposons. *Mol. Cell. Biol.* **6:** 168.

Lomidze, N.V., D.A. Kramerov, and A.P. Ryskov. 1986. Comparative analysis of

B1-like and B2-like repeated sequences in the DNA of different organisms. *Mol. Biol.* **20**: 612.

Nakamura, Y., M. Leppert, P. O'Connell, R. Wolff, T. Holm, M. Culver, C. Martin, E. Fujimoto, M. Hoff, E. Kumlin, and R. White. 1987. Variable number of tandem repeats (VNTR) markers for human gene mapping. *Science* **235**: 1616.

Nickerson, D.A., C. Whitehurst, C. Boysen, P. Charmley, R. Kaiser, and L. Hood. 1992. Identification of clusters of biallelic polymorphic sequence-tagged sites (pSTSs) that generate highly informative and automatable markers for genetic linkage mapping. *Genomics* **12**: 377.

Rogaev, E.I. 1990. Simple human DNA-repeats associated with genomic hypervariability, flanking the genomic retroposons and similar to retroviral sites. *Nucleic Acids Res.* **18**: 1879.

Shakhmuradov, I.A. and N.A. Kolchanov. 1989. tRNA as a possible primer for initiation of reverse transcription of the dispersed repeats Alu, B1, B2 and L1. *Mol. Biol.* **23**: 897.

Siemieniak, D.R. and J.L. Slightom. 1990. A library of 3342 useful nonamer primers for genome sequencing. *Gene* **96**: 121.

Siemieniak, D.R, L.C. Siev, and J.L. Slightom. 1991. Strategy and methods for directly sequencing cosmid clones. *Anal. Biochem.* **192**: 441.

Sulston, J. 1992. Reality of sequencing cost—Reply. *Nature* **357**: 106.

Uberbacker, E.C. and R.J. Mural. 1991. Locating protein coding region in human DNA sequences using a neural network—Multiple sensor approach. *Proc. Natl. Acad. Sci.* **88**: 11261.

Wilson, R.K., B.F. Koop, C. Chen, N. Halloran, R. Sciammis, and L. Hood. 1992. Nucleotide sequence analysis of the 95 kb 3' terminal region of the murine T-cell receptor α/δ chain locus: Strategy and methodology. *Genomics* **13**: 1198.

Wilson, R.K., A.S. Yuen, S.M. Clark, C. Spence, P. Arakelian, and L. Hood. 1988. Automation of dideoxynucleotide DNA sequencing reactions using a robotic workstation. *BioTechniques* **6**: 776.

Winoto, A. and D. Baltimore. 1989a. A novel, inducible and T-cell specific enhancer located at the 3' end of the T-cell receptor α locus. *EMBO J.* **8**: 729.

———. 1989b. α-β lineage specific expression of the α T-cell receptor gene by nearby silencers. *Cell* **59**: 649.

Construction of a 10-Mb Physical Map in the Adenomatous Polyposis Region of Chromosome 5

Ellen Solomon and Anna-Maria Frischauf

Imperial Cancer Research Fund
Lincoln's Inn Fields
London WC2A 3PX, United Kingdom

We have produced a physical map of about 10 Mb in the region of the adenomatous polyposis coli (APC) gene. The framework of this map came from three somatic cell hybrids, each with a chromosome 5 containing a unique deletion in the APC region. We attempted to cover the common region of deletion with probes isolated from irradiation hybrids, end cloning, and microdissection. The map has been assembled largely by pulsed field gel electrophoresis (PFGE). A few yeast artificial chromosomes (YACs) were also used to cover the region immediately adjacent to APC.

Main points discussed include:

- approaches to localizing tumor suppressor genes
- usefulness of constitutional cytogenetic abnormalities
- irradiation hybrids as a source of probes
- end cloning as a strategy for isolating probes
- map assembly by PFGE
- current strategies for physical mapping

INTRODUCTION

We have created a physical map of about 10 Mb of chromosome 5 as a result of a positional cloning effort, by ourselves and other workers, to isolate the APC gene. Mutations in this gene lead to a dominantly inherited form of cancer of the colon, known as familial adenomatous polyposis, or FAP. Although this disease itself is relatively rare, sporadic colon cancer occurs at a rate of 25,000 cases per year in the United Kingdom and results in 19,000 deaths annually. The intensive search for the APC gene was driven by the hypothesis that it would be found to be mutated not only constitutionally in FAP patients, but also in the somatic cells of sporadic colon tumors, and that the isolation and characterization of this gene would lead to fundamental biological understanding of a very common cancer.

The hypothesis that sporadic colon cancer and FAP were related by common genetic events came from theoretical considerations (Knudson 1985) and from the well-studied example of the rare childhood tumor, retinoblastoma (Knudson 1971; Cavenee et al. 1983). The model predicted that a germ-line mutation in the APC gene would lead to an inactive gene product from one allele and, in the FAP tumors, a second mutation in the normal allele would lead to total loss of function of this gene product. In sporadic tumors, mutation of both alleles would occur in the somatic tissue. If this were the case, the gene could be considered to be acting recessively at the cellular level and would be classified as a tumor suppressor gene.

The isolation of such a gene must begin with the identification of a relatively small chromosomal region encompassing the gene. In a case such as colon cancer where the disease exists in both an inherited and a sporadic form, initial clues as to this chromosomal region may be derived from several different approaches.

Most valuably, but rarely, a family with the disease also carries a cytogenetically visible chromosome anomaly, which segregates with this disease. This is usually the most unambiguous and quickest way to determine the general location of the disease gene, although the likelihood of finding such an anomaly is small. Fortunately, in the case of APC, an individual was identified who had multiple physical and mental anomalies and FAP and who carried a constitutional deletion in chromosome 5q15-q22 (Herrera et al. 1986). Assuming the cytogenetic event is straightforward and does not involve complex rearrangements, this type of deletion identifies the subchromosomal region in which the gene must lie. Unlike microscopically detectable deletions that cover rather large physical distances—of say 10 Mb—translocations may occur directly in the gene of interest and are therefore of even greater value. For example, the search for the von Recklinghausen neurofibromatosis (NF1) gene was aided enormously by the discovery of two separate translocations in different affected individuals, both involving the same small region of chromosome 17 (Rey et al. 1987; Ledbetter et al. 1989). In any gene search, karyotyping of one affected individual from each family can potentially result in an enormous reduction of time for relatively little effort.

The classic approach to the regional localization of any inherited syndrome, including those involving cancer, is genetic linkage. Increasingly tight coverage of the genome with highly polymorphic DNA markers has made this type of effort increasingly successful (Solomon and Bodmer 1979; Botstein et al. 1980). By now, 12 dominantly inherited cancer-predisposing loci have been localized by genetic linkage (Goddard and Solomon 1993). A sufficient number of large multigenerational families and well-spaced highly informative markers should in theory allow localization within 2 cM, although in practice this is rarely achieved. Once a distance of perhaps 3 or 4 cM is defined, further narrowing depends on the identification of key recombinant individuals within this region. In the absence of cytogenetic clues, the initial linkage analysis is a time-consuming, labor-intensive operation. The fortuitous discovery of an individual with a 5q deletion and FAP led immediately to linkage of this disease to probes close to this region of chromosome 5 (Bodmer et al. 1987; Leppert et al. 1987), although informative recombinants were too few to provide substantial reduction of the candidate region.

Unlike many positional cloning efforts, the search for tumor suppressor genes may employ an additional very powerful approach, which utilizes the sporadic tumors rather than family material. The mutational mechanisms by which a tumor suppressor gene can be deleted or inactivated range from small localized events within the gene, such as point mutations, to interstitial deletions and mitotic recombination, to loss of the whole chromosome. The mechanisms involving gross rearrangements such as deletions and chromosomal loss occur quite commonly, and the resulting loss of genetic material has become the hallmark of a tumor suppressor gene. Comparison of polymorphic markers on DNA digests of normal tissue with tumor tissue may show loss of one of the alleles in the tumor. If a whole chromosome is lost, allele loss will be seen with different markers along the whole chromosome, and regional information is not gained. However, if the mutational events are more localized, such as with interstitial deletions and mitotic recombination, allele loss can be an extremely valuable means of narrowing the boundaries of the region; it has been used successfully for DCC, a second tumor suppressor gene involved in colon cancer, on chromosome 18 (Fearon et al. 1990).

Finally, because the mutational mechanisms inactivating tumor suppressor genes can include submicroscopic deletions, further narrowing of a region defined by linkage or allele loss is possible if such deletions can be detected. Submicroscopic deletions cannot be systematically searched for until the region is quite narrow and covered with a fairly high density of probes. When this degree of localization has been achieved, rearrangements on PFGE blots, using affected family members and tumors, may reveal small deletions or rearrangements. Caution must be taken with this approach insofar as apparent changes on PFGs may be difficult to interpret and may be due to factors such as differences in methylation between cell lines. However, using a probe that showed high allele loss in sporadic tumors, alterations were detected on PFGs in two patients with FAP (Joslyn et al. 1991). These did indeed turn out to be due to small deletions of 100 kb and 260 kb, which not only helped narrow the region, but also ultimately helped identify the correct candidate APC gene.

A search for a tumor suppressor gene, then, must begin with a wide sweep, including cytogenetics on affected family members, linkage for flanking markers and key recombinants, and allele loss and PFGE studies in the search for interstitial deletions. The use of these approaches inevitably generates a large amount of genetic and physical map information in the surrounding region. We have produced a physical map of roughly 10 Mb around APC and describe here more specifically the approaches used.

Deletion hybrids

We were able to obtain material from three individuals with deletions in the same region of chromosome 5 as in the original FAP case mentioned above. The first was a young girl, M, aged 10, with mildly dysmorphic features and mild developmental delay, whose APC status was unknown (Varesco et al. 1989). Her deletion covered the region of 5q15-q22. The second individual, P, was one of two brothers affected with FAP whose mother died of colorectal cancer. These brothers were in their 40s, both with dysmorphic features and mental retardation. Again, their chromosomes 5 carried a 5q15-q22 deletion (Hockey et al. 1989). The third was a case, S, with FAP, dysmorphic features, and mental retardation in whom the 5q deletion, described as 5q22.1-q23.2, was due to an intrachromosomal insertion and in which the distal boundary of the deletion appeared considerably more telomeric than in the other two cases (Cross et al. 1992).

Somatic cell hybrids were made by fusing temperature-sensitive Chinese hamster cells with lymphoid lines established from these three patients, M, P, and S. Growth at high temperature disables the Chinese hamster leucyl-tRNA synthetase, and the human gene product is therefore required for survival of the hybrids (Thomson et al. 1973; Giles et al. 1980). The human leucyl-tRNA synthetase gene is on chromosome 5, which is therefore selected at the high temperature. From these fusions we obtained a hybrid with a normal human chromosome 5, PN/TS-1, and no other human material, as well as hybrids with each of the 5del chromosomes, PD/TS-1, MD/TS-1, and SD/TS-1, with little or no other human material. These 4 hybrids produced a panel into which probes could be mapped, as shown in Figure 1. In panel A, a cosmid isolated with a microdissected clone (see below, Microcloning) from the FAP region is hybridized to a normal chromosome 5 (PN/TS-1) and the three deletion hybrids. The clone is deleted in all three and therefore lies within the region of interest. In panel B, clone 12.25, from irradiation hybrid 12 (see below, Irradiation hybrids) is hybridized to the same four hybrids. It is deleted in MD/TS-1 and SD/TS-1 but not PD/TS-1, indicating that the PD deletion is smaller than the others at one boundary. This is the sole clone that distinguishes the PD and SD distal breakpoints, as

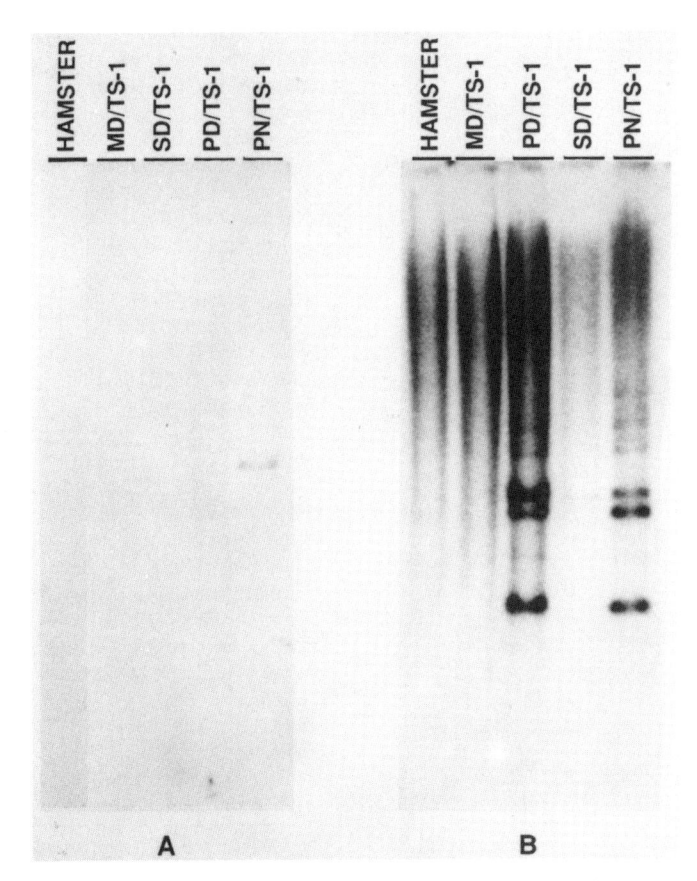

Figure 1 DNA from hamster line tsH1 and hybrids containing a normal chromosome 5 (PN/TS-1) and chromosome 5 dels (PD/TS-1, SD/TS-1, MD/TS-1). (A) DNA digested with *Bgl*II, probe cMC411Bam1.0. (B) DNA digested with *Bgl*II, probe 12.25Hind3.1.

Table 1 Probes used in the construction of the physical map

Probe	S	P	M	Source	Locus	Reference
C11P11	+	+	+		D5S71	Bodmer et al. (1987)
KK5.33	+	–	–	R. White	D5S85	Nakamura et al. (1988)
14.20Sal/Hind	+	–	–	a		Ward et al. (1993)
12.3Eco5	+	–	–	a		Ward et al. (1993)
27F1Sal/Sst	+	–	–	b		Ward et al. (1993)
ECB27Sal2.8	+	–	–	c	D5S98	Varesco et al. (1989)
pYN5.64Eco1.0	+	–	–	R. White	D5S82	Nakamura et al. (1988)
FER	–	–	–	B. Vogelstein		Morris et al. (1990)
pCB83.6	–	–	–	Meera Khan	D5S122	Breukel et al. (1991)
220F1Nco/Bam	–	–	–	b		Ward et al. (1993)
ECB220Eco2.4	–	–	–	c	D5S114	Varesco et al. (1989)
L5.62	–	–	–	B. Vogelstein	D5S134	Kinzler et al. (1991b)
12.75Eco3.0	–	–	–	a		Ward et al. (1993)
TB1	–	–	–	B. Vogelstein		Kinzler et al. (1991b)
5.3.1.1	–	–	–	d 89,8G	D5S106	Ward et al. (1993)
cMC534Eco3.1	–	–	–	e 168,12A		Hampton et al. (1991a)
cMC241Eco3.2	–	–	–	e		Hampton et al. (1991a)
EF5.44	–	–	–	B. Vogelstein	D5S135	Kinzler et al. (1991b)
yA1010.LE	–	–	–	f 66,5D		Hampton (1992)
FB9A	–	–	–	B. Vogelstein		Kinzler et al. (1991b)
YM72	–	–	–	g 180,8E		Hampton et al. (1992)
FB70	–	–	–	B. Vogelstein		Kinzler et al. (1991b)
TB2	–	–	–	B. Vogelstein		Kinzler et al. (1991b)
YM8 Eco 1.9	–	–	–	g 15,3F		Hampton et al. (1992)
SW15	–	–	–	B. Vogelstein		Kinzler et al. (1991a)
YM30Eco/BssH0.5	–	–	–	g 26,8E		Ward et al. (1993)
YM30Eco/BssH2.5	–	–	–	g 26,8E		Ward et al. (1993)
YM39Bam1.0	–	–	–	g 17,3E		Ward et al. (1993)
YM39Pst	–	–	–	g 17,3E		Ward et al. (1993)
0624RE	–	–	–	f		Hampton (1992)
14.16Bgl1.9	–	–	–	a		Ward et al. (1993)
MC213Eco2.4	–	–	–	d 25,10B		Hampton et al. (1991a)
YN5.48	–	–	–	R. White	D5S81	Nakamura et al. (1988)
14.17	–	–	–	a		J.R.T. Ward (unpubl.)
G016RE	–	–	–	f		Hampton (1992)
MC451Eco0.6	–	–	–	d 21,7A		Hampton et al. (1991a)
FAP14	–	–	–	h	D5S327	Hampton et al. (1991a)
MC411Bam1.0	–	–	–	d 13,6A		Hampton et al. (1991b)
12.46Bam3.0	–	–	–	a		Ward et al. (1993)
12.25Hind3.1	–	+	–	a		Ward et al. (1993)
ECB134Sst1.1	+	+	–	c	D5S97	Varesco et al. (1989)
14.61Bgl2.3	+	+	–	a		Ward et al. (1993)
MC5.61	+	+	–	R. White	D5S84	Nakamura et al. (1988)
EW5.5	+	+	–	R. White	D5S86	Nakamura et al. (1988)

S, P, and M are the deletion Chr. 5 as described in text. (+) Marker present in deleted chromosome; (–) marker absent. Sources: (a) Fragments of λ clones from Chr. 5 irradiation hybrids. (b) Fragments of λ clones obtained by walking from a *Bss*HII end clone. (c) Subclones λ end clones. (d) Fragments of cosmids from the Chr. 5 library screened with microclones. Coordinates are given. (e) c5.3.1 was isolated from the cosmid library using p3.1 (Feder et al. 1985) (obtained from N. Spurr) as a probe. p3.1 hybridized to hamster DNA and a human locus not on Chr. 5. Subclones of c5.3.1 used for PFG mapping hybridized to human Chr. 5 only. (f) Probes derived from the ends of YACs. (g) Subclones or fragments derived from screening the Chr. 5 cosmid library with YACs B0624 and A1010. Coordinates are given. (h) Microclone.

shown in Table 1. The original marker, C11P11, to which FAP was linked at 15% recombination, was clearly shown to be outside the deletions. Using this panel, we were able to map all probes available from the region. Additionally, with the MD/TS-1 hybrid, containing the deletion 5, we were able to show that this deletion included within it the flanking markers for APC, suggesting that the individual is at high risk for the disease. In this way we were able to provide the first counseling for this disease based on molecular genetic information.

Table 1 shows the markers mapped with respect to the deletions and the relative positions of the deletions with respect to each other. Clearly the S deletion, although appearing further distal cytogenetically, has a very similar distal boundary to P. A variety of methods, described below, were used to isolate further probes with which to cover the smallest common region of the deletions. This hybrid mapping panel was used for mapping all of these probes and formed a framework for the physical map.

Irradiation hybrids

Cox et al. (1990), using a modification of the technique of Goss and Harris (1975), demonstrated that irradiation of a somatic-cell hybrid containing a single human chromosome, followed by fusion to a rodent line, resulted in retention of unselected fragments of the human chromosome. For chromosome 21, with a dose of 8000 rads, the retention of markers was 30–60% (Cox et al. 1990). With chromosome 11, a dose of 9000 rads resulted in an average retention of markers of 26% (Richard 1991). The hybrids could be used successfully as a mapping tool to determine both the distance between probes and the order of probes on the chromosome. By analysis of the pairwise retention of markers, the distance between them could be estimated. A best order of all the probes could be determined by analyzing the minimal number of breaks. In addition, it was found that the fragments could be used as an enriched source of markers for particular regions.

We decided to make an irradiation hybrid panel using the single chromosome hybrid, PN/TS-1, containing a normal 5 as its only human material. This was to be primarily a means of generating probes from the APC region, but the possibility of obtaining order of the probes using this panel was also considered. We chose a high dose of irradiation, 50,000 rad, with the expectation that the fragments would be very small and highly specific for the region of interest (Benham et al. 1989). From the fusion, 150 independent colonies were picked. Analysis of the *Alu* polymerase chain reaction (PCR) products demonstrated that 100% of the colonies contained human material. Scoring for markers along the long arm of chromosome 5 indicated an average retention of markers of 9%, considerably lower than had been seen with lower doses of irradiation. In situ hybridization of several of the clones with human DNA showed one or two fragments. A subpanel was made with those hybrids positive for ECB27, a marker within the deletions.

Nine of the hybrids that were positive for either ECB27 or YN5.48 (a second marker within the deletions) were made into phage libraries as a source of probes. These were mapped onto the panel of deletion hybrids with the expectation that they would be enriched for this region. A total of 201 clones were mapped, of which 10 fell within the deletions. The numbers from the individual libraries were 0/8 (hybrid 4); 4/47 (hybrid 12); 0/17 (hybrid 13); 5/22 (hybrid 14); 0/18 (hybrid 32); 1/25 (hybrid 58); 0/18 (hybrid 115); 0/30 (hybrid 137); 0/16 (hybrid 158). This number of 5% is approximately what would be expected by mapping random clones into the deletions using a whole chromosome 5 library. Although no enrichment was obtained, the effort did result in a number of useful probes in the region. These are designated with the prefixes 14.- and 12.- and are shown in Table 1 and Figure 2. In situ hybridization with pooled phage clones from three libraries showed scattering of signal over the whole chromosome, indicating that the blocks of inserted human DNA were composed of material from multiple regions of the chromosome. PFGE of several hybrids probed with ECB27 gave band sizes quite different from those of normal DNA and indicated that on average the size of the fragments was less than 1 Mb. It could be estimated that these hybrids contained as many as 20 very small fragments from different parts of the chromosome, which, at least in some of them, appeared to be inserted in one or two blocks. Ordering of markers with the small fragments was unsuccessful, as markers within the deletion region were not associated in pairwise combinations at frequencies higher than random, even when they were known to be very close. Ultimately, two markers from within the APC gene were tested, and these also did not cosegregate.

We concluded that although we did produce probes for the region, a much lower dose of irradiation would have been required to have made this a really efficient strategy.

End cloning

Long-range restriction mapping around C11P11, the first marker shown to be linked to APC, indicated that the region was relatively poor in CpG-rich islands (Bird 1987). We therefore decided not to construct a linking

Figure 2 Map segment containing the APC gene. The order of the markers was established by pulsed field gel electrophoresis (Ward et al. 1993). Only the unmethylated or partially unmethylated *Mlu*I (M) sites present in PN(TS) DNA are given here. B shows the *Bss*HII sites for which end clones were derived (section 3). References for the probes are given in Table 1. Underlined probes are polymorphic and have been used in linkage and allele loss studies.

PHYSICAL MAP OF CHROMOSOME 5   **97**

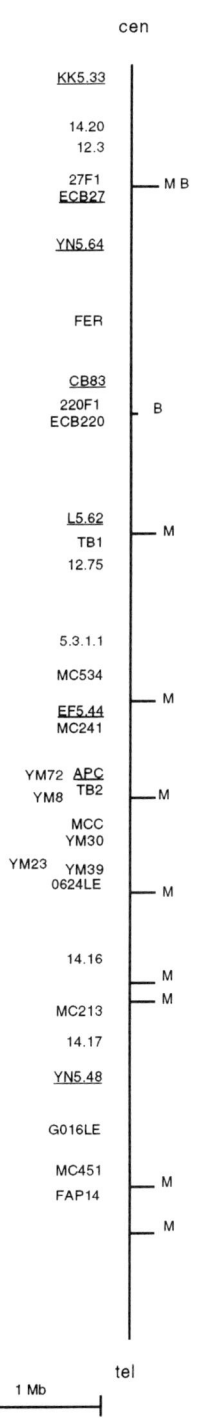

```
                cen
    KK5.33       |
     14.20       |
     12.3        |
     27F1        |— M B
     ECB27       |
                 |
    YN5.64       |
                 |
                 |
     FER         |
                 |
     CB83        |
     220F1       |. B
     ECB220      |
                 |
                 |
     L5.62       |
     TB1         |— M
     12.75       |
                 |
                 |
    5.3.1.1      |
    MC534        |
    EF5.44       |— M
    MC241        |
                 |
  YM72  APC      |
  YM8   TB2      |— M
                 |
        MCC      |
        YM30     |
 YM23   YM39     |
        0624LE   |— M
                 |
                 |
     14.16       |
                 |— M
                 |— M
    MC213        |
     14.17       |
    YN5.48       |
                 |
    G016LE       |
                 |
    MC451        |
    FAP14        |— M
                 |
                 |— M
                 |
                 |
                tel
  |—1 Mb—|
```

Figure 2 (*See facing page for legend.*)

library for *Not*I or another rare-cutting enzyme, since these libraries are enriched for clones from CpG-island-rich regions. There are, however, several advantages to the use of clones containing the sites for rare-cutting enzymes in the construction of long-range restriction maps, as has been discussed previously (Frischauf 1989): Linking clones can be accurately localized on the map since they cover the site for the restriction enzyme; they hybridize to the fragments on both sides of the site; they avoid the fortuitous isolation of multiple probes that all characterize the same restriction fragment; and, if they have been obtained with enzymes that typically cut within CpG-rich islands, they are usually associated with the 5′ ends of genes. To retain these advantages, we took the approach of isolating long restriction fragments by PFGE and cloning their ends (Michiels et al. 1987). This technique results in the generation of "half linking clones" that can easily be converted to full linking clones by screening of conventional phage or cosmid libraries. The additional purification step of size fractionation eliminates most of the clones from island-rich regions, since they would be derived from relatively short fragments. Whether size fractionation and end cloning constitute an efficient approach is determined by the availability of reduced hybrids. Total human genomic DNA cannot be used in such a strategy, since size fractionation on a gel usually results in no more than a 10- to 50-fold enrichment on a mass basis, but a monochromosomal hybrid can give a good yield of clones from the right region. On the other hand, if a sufficiently reduced hybrid with no CpG-island-rich regions were available, an end cloning approach would be unnecessarily laborious.

When we started to isolate new markers from the 5q21-q23 region, the only hybrid available to us was H64 (MacDonald et al. 1987) containing human chromosomes 4 and 5 on a hamster background. We knew that C11P11, the marker first shown to be linked to APC, was localized in an island-poor region on a *Bss*HII fragment greater than 800 kb. We therefore chose to construct a library of end clones from long *Bss*HII fragments to isolate markers for restriction mapping. We obtained the ends of 0.8-Mb to 1.5-Mb *Bss*HII fragments from H64 DNA and mapped them to the 5q21-23 region using the hybrids PD/TS-1 and MD/TS-1. As expected, the total number of clones from the region of interest was small. Of 147 clones, only 2 mapped to the PD and MD deletions, and 1 further clone mapped to the region deleted only in MD (Varesco et al. 1989). The 2 former clones were, however, essential in anchoring and orienting the whole proximal part of the map where markers are relatively sparse and fragments are long. One of the *Bss*HII end clones also is an *Mlu*I end clone, and after walking across the rare-cutting sites in a phage library, those two marker clones "see" a total of about 5 Mb of DNA in *Mlu*I and *Bss*HII fragments in the cell line GM1416B. End cloning can thus be an efficient approach to isolating markers for long-range restriction mapping.

Microcloning

As a complementary approach, direct cutting out of bands from lightly fixed chromosomes using a micromanipulator has been shown to be a very efficient way of isolating DNA from a specific chromosomal region (Scalenghe et al. 1981; Rohme et al. 1984; Ludecke et al. 1990). We obtained microclones from the 5q22 region that were mapped back to the region by hybridization to the hybrid panel containing the deletion chromosomes 5 (Fig. 1) (Hampton et al. 1991a,b). The clones either were used directly as PFGE probes or were hybridized to an arrayed chromosome 5 cosmid library (obtained from L. Deaven, Los Alamos) and the isolated cosmids were again checked against the hybrid panel. Unique fragments were then isolated and used as probes for the long-range restriction map. As these probes were placed on the map, their distribution was shown not to be entirely random, since two microclones were shown to be located on the same cosmid, whereas there remained large regions from which no microclones had been isolated (J.R.T. Ward and G.M. Hampton, unpubl.). Overall, however, the microclones were an excellent source of markers well distributed over the region.

PFGE

PFGE is the established method for producing long-range restriction maps of megabase segments of DNA. We have primarily relied on PFGE in constructing the 5q21-q23 map (Ward et al. 1993). The density of markers is one critical parameter in obtaining a reliable map; the other is the density of CpG islands. When CpG islands are closely spaced and contain sites for most of the enzymes normally used in map construction, then a high density of markers per length of DNA is required and enzymes that typically cut outside CpG islands, like *Cla*I, *Sfi*I, and *Nru*I (Brown and Bird 1986), have to be included in the analysis to find enough bridging fragments. When CpG islands are far apart, a lower density of markers is necessary since the restriction fragments are long, but it may be difficult to resolve them from each other and from the "limiting mobility" band where all DNA above a certain size will travel under a given set of conditions. Errors in size determination of large bands are proportionately larger than for small fragments. Fragment patterns derived from cutting the rare restriction sites between islands are often complicated, since those sites can be partially methylated. This leads to a larger number of bands with unknown overlaps that are difficult to interpret.

For the construction of the long-range restriction map of 5q21-23, we decided to focus on ordering the markers, giving a lower priority to the exact determination of distances. To limit the amount of work involved in mapping a 10-Mb region, we chose a moderate number of restriction enzymes with different, complementary properties. *Not*I,

*Bss*HII, and *Eag*I typically cut in CpG-rich islands. The latter two cut much more frequently than *Not*I and are therefore suitable for the analysis of highly methylated regions where the *Not*I fragments are too large to be resolved. *Not*I is used to cut the more CpG-island-rich segments into manageable pieces. Sites for the fourth main enzyme, *Mlu*I, are much less restricted to CpG islands and often allow the connection of markers across them. These four enzymes were occasionally insufficient to connect groups of markers and were supplemented by *Nru*I or *Sfi*I.

We primarily connected probes by showing that they hybridized to fragments of identical mobility on digestion with several different restriction enzymes. When we could find only one enzyme that gave such a fragment, we took advantage of "methylation polymorphism" to show that both probes hybridized to the same rather than two comigrating different restriction fragments (Bucan et al. 1990). This strategy is based on possible differences in the methylation status of restriction sites for methylation-sensitive rare-cutting enzymes in different cell lines or blood samples. If a restriction site between two probes is methylated in some DNA samples and not in others, then these two probes will hybridize to the same fragment in a subset of the samples. Where the site is unmethylated, the two probes will hybridize to different, smaller restriction fragments and the sizes of those two fragments will usually add up to the longer fragment that is seen in the first subset. If the different hybridization patterns of the two probes are due to presence or absence of methylation of the same restriction site, then the pattern always has to change for both at the same time. If this condition is fulfilled, the probability of random comigration of hybridizing fragments is very much reduced and it can be assumed that the two probes are physically connected.

We have frequently relied on this method in the construction of the long-range restriction map. Sometimes we have connected two probes across a CpG island containing many rare-cutting restriction sites by showing that they hybridized to the same YAC clone. The probes are listed in Table 1, and the map segment containing the APC gene is shown in Figure 2. The map spans 10 Mb using 36 probes at an average density of 1 probe per 280 kb. The probes are not evenly distributed and are at their highest concentration in the region containing the APC and MCC genes, reflecting both the focus of interest and the necessity for more probes due to a higher density of genes with CpG-rich islands (Lindsay and Bird 1987). Connecting the more proximal part of the map to the APC/MCC region by PFGE indicated the order cen-APC-MCC-tel, in contrast to data of Kinzler et al. (1991b). Because of the importance of this region, we used in situ hybridization on interphase nuclei to confirm the orientation of this map segment, which is otherwise completely compatible with the published data (Kinzler et al. 1991b; J.R.T. Ward et al., in prep.). A separate segment of the map corresponding to 3 Mb located more distally could not be connected to the main body of the map. This fragment is interesting in that it contains the breakpoint of two of the deletions isolated in hybrids and used for mapping. The distal breakpoints of **PD** and **PS** are less than 900 kb apart, a surprising finding considering the different cytological appearance (see above, Deletion hybrids). Despite the large number of markers that were used in the construction of the map, some of the connections would have been easier if we had relied more extensively on YACs and their ends in our analysis. We did, however, make limited use of this technique, as described below.

YAC clones

Several YAC clones were isolated from the ICRF library (Larin et al. 1991) using microclones that mapped within the smallest of the three deletions, PS, used in regional mapping. The region surrounding the APC and MCC genes was isolated using an MCC cDNA probe (Kinzler et al. 1991a; Hampton et al. 1992). Whereas the former YACs mainly served to help connect probes thought to be physically close but separated by strong CpG islands, the latter were converted to cosmids covering the region of interest. This was done by excising the artificial chromosome from pulsed field gels, labeling, and hybridizing to an arrayed chromosome 5 cosmid library spotted on filters at high density (Nizetic et al. 1991). Cosmids giving a positive hybridization signal in the presence of a large amount of Cot1 DNA can be considered subclones of the YAC (Baxendale et al. 1991). An additional advantage of this procedure is that when chimeric YACs are used as probes, only cosmids corresponding to the chromosome 5 portion are obtained. We also used an end probe of YAC, 0624RE to connect the APC region to the more distal markers (Ward et al. 1993).

DISCUSSION AND OUTLOOK

We have saturated the 5q21-23 region with DNA markers and used these to construct a physical map of the region, predominantly by restriction mapping with rare-cutting enzymes and PFGE. In our search for new markers, we isolated end clones from a hybrid containing chromosomes 4 and 5, and random λ clones from hybrids carrying irradiation-induced fragment of chromosome 5. We relied on hybrids carrying chromosomes 5 deleted for the region of interest to assign new probes to that region. At a late stage in the marker search, we obtained microclones from the region we wanted to map. These were used either directly in hybridization to PFGE blots, or after corresponding cosmids had been isolated. Some segments are covered by YAC clones.

The project as summarized here was carried out over an extended period of time in which there was considerable development in the techniques available for large-scale mapping and cloning. Although the basic strategy of first saturating a region with markers and then trying to connect them is still valid, a comparable project initiated now might replace much PFGE analysis with a combination of YAC cloning and in situ hybridization to order the YAC contigs. Deletions associated with the phenotype would still be invaluable, and clones from radiation hybrids or microclones can provide multiple starting points for isolating YACs.

Very little use was made of microsatellite markers (Litt and Luty 1989; Smeets et al. 1989; Weber and May 1989), a major development during the search for the APC gene. This is true not only for our group, but also for most of the other groups involved in this project. This partly reflects the scarcity of unambiguous recombinants to be analyzed in detail, the isolation and distribution of polymorphic cosmid markers by Nakamura (Nakamura et al. 1988), and the difficulties in using PCR-based markers for allele loss studies. Most of the markers that we developed were aimed at first establishing a physical distance for the flanking markers, and we made no effort to look for polymorphism. Given the relative ease of obtaining microsatellite markers from YAC clones, this approach would certainly have a much higher priority in the early stages of the search if it was initiated now. Further changes in the most efficient approach to identifying genes will undoubtedly come about as ordered YAC and cosmid contigs become available for larger areas of the genome (Chumakov et al. 1992). Physical mapping should cease to require such a major effort and the resources will be devoted to the preceding and the following steps: the genetic analysis of interesting genes and the testing of candidates from the mapped region.

Acknowledgments

We thank J.R. Tristan Ward and Huw Thomas for unpublished data.

References

Baxendale, S., G.P. Bates, M.E. MacDonald, J.F. Gusella, and H. Lehrach. 1991. The direct screening of cosmid libraries with YAC clones. *Nucleic Acids Res* **19**: 6651.

Benham, F., K. Hart, J. Crolla, M. Bobrow, M. Francavilla, and P. N. Goodfellow. 1989. A method for generating hybrids containing nonselected fragments of human chromosomes. *Genomics* **4**: 509.

Bird, A.P. 1987. CpG islands as gene markers in the vertebrate nucleus. *Trends Genet.* **3**: 342.

Bodmer, W.F., C.J. Bailey, J. Bodmer, H.J.R. Bussey, A. Ellis, P. Gorman, F.C.

Lucibello, V.A. Murday, S.H. Rider, P. Scambler, D. Sheer, E. Solomon, and N.K. Spurr. 1987. Localization of the gene for familial adenomatous polyposis on chromosome 5. *Nature* **328**: 614.

Botstein, D., R.L. White, M. Skolnick, and R.W. Davis. 1980. Construction of a genetic linkage map in man using restriction fragment length polymorphisms. *Am. J. Hum. Genet.* **32**: 314.

Breukel, C., C. Tops, E. Ras, and P.M. Khan. 1991. Mspl RFLP at the D5S122 locus tightly linked to APC. *Nucleic Acids Res.* **19**: 685.

Brown, W.R. and A.P. Bird. 1986. Long-range restriction site mapping of mammalian genomic DNA. *Nature* **322**: 477.

Bucan, M., M. Zimmer, W.L. Whaley, A. Poustka, S. Youngman, B.A. Allitto, E. Ormondroyd, B. Smith, T.M. Pohl, M. MacDonald, G.P. Bates, J. Richards, S. Volinia, T.C. Gilliam, Z. Sedlacek, F.S. Collins, J.J. Wasmuth, D.J. Shaw, J.F. Gusella, A.-M. Frischauf, and H. Lehrach. 1990. Physical maps of 4p16.3, the area expected to contain the Huntington disease mutation. *Genomics* **6**: 1.

Cavenee, W.K., T.P. Dryja, R.A. Phillips, W.F. Benedict, R. Godbout, B.L. Gallie, A.L. Murphree, L.C. Strong, and R.L. White. 1983. Expression of recessive alleles by chromosomal mechanisms in retinoblastoma. *Nature* **305**: 779.

Chumakov, I., P. Rigault, S. Guillou, P. Ougen, A. Billaut, G. Guasconi, P. Gervy, I. LeGall, P. Soularue, L. Grinas, L. Bougueleret, C. Bellanne-Chantelot, B. Lacroix, E. Barillot, P. Gesnouin, S. Pook, G. Vaysseix, G. Frelat, A. Schmitz, J.-L. Sambucy, A. Bosch, X. Estivill, J. Weissenbach, A. Vignal, H. Riethman, D. Cox, D. Patterson, K. Gardiner, M. Hattori, Y. Sakaki, H. Ichikawa, M. Ohki, D. Le Paslier, R. Heilig, S. Antonarakis, and D. Cohen. 1992. Continuum of overlapping clones spanning the entire human chromosome 21q [see comments]. *Nature* **359**: 380.

Cox, D.R., M. Burmeister, E.R. Price, S. Kim, and R.M. Myers. 1990. Radiation hybrid mapping: A somatic cell genetic method for constructing high-resolution maps of mammalian chromosomes. *Science* **250**: 245.

Cross, I., J. Delhanty, P. Chapman, L.V. Bowles, D. Griffin, J. Wolstenholme, M. Bradburn, J. Brown, C. Wood, A. Gunn, and J. Burn. 1992. An intrachromosomal insertion causing 5q22 deletion and familial adenomatous polyposis coli in two generations. *J. Med. Genet.* **29**: 175.

Fearon, E.R., K.R. Cho, J.M. Nigro, S.E. Kern, J.W. Simons, J.M. Ruppert, S.R. Hamilton, A.C. Preisinger, G. Thomas, K.W. Kinzler, and B. Vogelstein. 1990. Identification of a chromosome 18q gene that is altered in colorectal cancers. *Science* **247**: 49.

Feder, J., H.M. Gurling, J. Darby, and L.L. Cavalli-Sforza. 1985. DNA restriction fragment analysis of the proopiomalanocortin gene. *Am. J. Hum. Genet.* **37**: 286.

Frischauf, A.-M. 1989. Construction and use of linking libraries. *Technique* **1**: 3.

Giles, R.E., N. Shimizu, and F.H. Ruddle. 1980. Assignment of a human genetic locus to chromosome 5 which corrects the heat sensitive lesion associated with reduced leucyl-tRNA synthetase activity in ts025C1 Chinese hamster cells. *Somatic Cell Genet.* **6**: 667.

Goddard, A.D. and E.S. Solomon. 1993. Genetic aspects of cancer. *Annu. Rev. Hum. Genet.* **21**: 319.

Goss, S.J. and H. Harris. 1975. New method for mapping genes in human chromosomes. *Nature* **255**: 680.

Hampton, G. 1992. "Molecular genetic analysis of the adenomatous polyposis coli (APC) gene region." Ph.D. thesis, University College, London.

Hampton, G.M., C. Howe, G. Leuteritz, H. Thomas, W.F. Bodmer, E. Solomon, and W.G. Ballhausen. 1991a. Regional mapping of 22 microclones around the adenomatous polyposis coli (APC) locus on chromosome 5q. *Hum. Genet.* **88:** 112.

Hampton, G.M., J.R.T. Ward, S. Cottrell, K. Howe, H.J.W. Thomas, W.G. Ballhausen, T. Jones, D. Sheer, E. Solomon, A.-M. Frischauf, and W.F. Bodmer. 1992. Yeast artificial chromosomes for the molecular analysis of the familial polyposis APC gene region. *Proc. Natl. Acad. Sci.* **89:** 8249.

Hampton, G., G. Leuteritz, H.J. Ludecke, G. Senger, U. Trautmann, H. Thomas, E. Solomon, W.F. Bodmer, B. Horsthemke, U. Claussen, and W.G. Ballhausen. 1991b. Characterization and mapping of microdissected genomic clones from the adenomatous polyposis coli (APC) region. *Genomics* **11:** 247.

Herrera, L., S. Katati, L. Gibas, E. Pietrzak, and A.A. Sandberg. 1986. Gardner syndrome in a man with an interstitial deletion of 5q. *Am. J. Med. Genet* **25:** 473.

Hockey, K.A., M.T. Mulcahy, P. Montgomery, and S. Levitt. 1989. Deletion of chromosome 5q and familial adenomatous polyposis. *J. Med. Genet.* **26:** 61.

Joslyn, G., M. Carlson, A. Thliveris, H. Albertsen, L. Gelbert, W. Samovitz, J. Groden, J. Stevens, L. Spirio, M. Robertson, L. Sargeant, K. Krapcho, E. Wolff, R. Burt, J.P. Hughes, J. Warrington, J. McPherson, J. Wasmuth, D. Le Paslier, H. Abderrahim, D. Cohen, M. Leppert, and R. White. 1991. Identification of deletion mutations and three new genes at the familial polyposis locus. *Cell* **66:** 601.

Kinzler, K.W., M.C. Nilbert, B. Vogelstein, T.M. Bryan, D.B. Levy, K.J. Smith, A.C. Preisinger, S.R. Hamilton, P. Hedge, A. Markham, A. Markham, M. Carlson, G. Joslyn, J. Groden, R. White, Y. Miki, Y. Miyoshi, I. Nishisho, and Y. Nakamura. 1991a. Identification of a gene located at chromosome 5q21 that is mutated in colorectal cancers. *Science* **251:** 1366.

Kinzler, K.W., M.C. Nilbert, L.K. Su, B. Vogelstein, T.M. Bryan, D.B. Levy, K.J. Smith, A.C. Preisinger, P. Hedge, D. McKechnie, R. Finniear, A. Markham, J. Groffen, M.S. Bogushi, S.F. Altschul, A. Horii, H. Ando, Y. Miyoshi, Y. Miki, I. Nishisho, and Y. Nakamura. 1991b. Identification of FAP locus genes from chromosome 5q21. *Science* **253:** 661.

Knudson, A.G. 1971. Mutation and cancer: Statistical study of retinoblastoma. *Proc. Nat. Acad. Sci.* **68:** 820.

Knudson, A.J. 1985. Hereditary cancer, oncogenes, and antioncogenes. *Cancer Res.* **45:** 1437.

Larin, Z., A.P. Monaco, and H. Lehrach. 1991. Yeast artificial chromosome libraries containing large inserts from mouse and human DNA. *Proc. Natl. Acad. Sci.* **88:** 4123.

Ledbetter, D.H., D.C. Rich, P. O'Connell, M. Leppert, and J.C. Carey. 1989. Precise localization of NF1 to 17q11.2 by balanced translocation. *Am. J. Hum. Genet.* **44:** 20.

Leppert, M., M. Dobbs, P. Scambler, P. O'Connel, Y. Nakamura, D. Stauffer, S. Woodward, R. Burt, J. Hughes, E. Gardner, M. Lathrop, J. Wasmuth, M. Lalouel, and R. White. 1987. The gene for familial polyposis coli maps to

the long arm of chromosome 5. *Science* **238**: 1411.
Lindsay, S. and A.P. Bird. 1987. Use of restriction enzymes to detect potential gene sequences in mammalian DNA. *Nature* **327**: 336.
Litt, M. and J.A. Luty. 1989. A hypervariable microsatellite revealed by in vitro amplification of a dinucleotide repeat within the cardiac muscle actin gene. *Am. J. Hum. Genet.* **44**: 397.
Ludecke, H.J., G. Senger, U. Claussen, and B. Horsthemke. 1990. Construction and characterization of band-specific DNA libraries. *Hum. Genet.* **84**: 512.
MacDonald, M.E., M.A. Anderson, T.C. Gilliam, L. Traneblaerg, N.J. Carpenter, E. Magenis, M.R. Hayden, S.T. Healey, T.I. Bonner, and J.F. Gusella. 1987. A somatic cell hybrid panel for localizing DNA segments near the Huntington's disease gene. *Genomics* **1**: 29.
Michiels, F., M. Burmeister, and H. Lehrach. 1987. Derivation of clones close to the cystic fibrosis marker met by preparative field inversion gel electrophoresis. *Science* **236**: 1305.
Morris, C., N. Heisterkamp, Q.L. Hao, J.R. Testa, and J. Groffen. 1990. The human tyrosine kinase gene (FER) maps to chromosome 5 and is deleted in myeloid leukemias with a del(5q). *Cytogenet. Cell Genet.* **53**: 196.
Nakamura, K., M. Lathrop, M. Leppert, M. Dobbs, J. Wasmuth, E. Wolff, M. Carlson, E. Fujimoto, K. Krapcho, T. Sears, S. Woodward, J. Hughes, R. Burt, E. Gardner, J.-M. Lalouel, and R. White. 1988. Localization of the genetic defect in familial adenomatous polyposis within a small region of chromosome 5. *Am. J. Hum. Genet.* **43**: 638.
Nizetic, D., G. Zehetner, A.P. Monaco, L. Gellen, B.D. Young, and H. Lehrach. 1991. Construction, arraying, and high-density screening of large insert libraries of human chromosomes X and 21: Their potential use as reference libraries. *Proc. Natl. Acad. Sci.* **88**: 3233.
Rey, J.A., M.J. Bello, J.M. de-Campos, J. Benitez, J.L. Sarasa, J.R. Boixados, and A. Sanchez-Cascos. 1987. Cytogenetic clones in a recurrent neuro-fibroma. *Cancer Genet. Cytogenet.* **26**: 157.
Richard, C.W., III., D.A. Withers, T.C. Meeker, S. Maurer, G.A. Evans, R.M. Myers, and D.R. Cox. 1991. A radiation hybrid map of the proximal long arm of chromosome 11 containing the multiple endocrine neoplasia type I (MEN-I) and bcl-1 disease loci. *Am. J. Hum. Genet.* **49**: 1189.
Rohme, D., H. Fox, B. Herrmann, A.-M. Frischauf, J.-E. Edstrom, P. Mains, L. Silver, and H. Lehrach. 1984. Molecular clones of the mouse t complex derived from microdissected metaphase chromosomes. *Cell* **36**: 783.
Scalenghe, F., E. Turco, J.-E. Edstroem, V. Pirotta, and M. Melli. 1981. Microdissection and cloning of DNA from a specific region of *Drosophila melanogaster* polytene chromosomes. *Chromosoma* **82**: 205.
Smeets, H.J.M., H.G. Brunner, H.-H. Ropers, and B. Wieringa. 1989. Use of variable simple sequence motifs as genetic markers application to study of myotonic dystrophy. *Hum. Genet.* **83**: 245.
Solomon, E. and W.F. Bodmer. 1979. Evolution of sickle variant gene. *Lancet* **I**: 923.
Thomson, L.H., L. Harkins, and C.P. Stanner. 1973. A mammalian cell mutant with a temperature sensitive leucyl-transfer RNA synthtase. *Proc. Natl. Acad. Sci.* **70**: 3094.
Varesco, L., H.J.W. Thomas, S. Cottrell, V. Murday, S.J. Fennell, S. Williams, S. Searle, D. Sheer, W.F. Bodmer, A.-M. Frischauf, and E. Solomon. 1989.

CpG island clones from a deletion encompassing the gene for adenomatous polyposis coli. *Proc. Natl. Acad. Sci.* **86:** 10118.

Ward, J.R.T., S. Cottrell, H.J.W. Thomas, T.A. Jones, C.M. Howe, G.M. Hampton, D. Sheer, W.F. Bodmer, E. Solomon, and A.-M. Frischauf. 1993. A long range restriction map of human chromosome 5q21-23. *Genomics* (in press).

Weber, J.L. and P.E. May. 1989. Abundant class of human DNA polymorphisms which can be typed using the polymerase chain reaction. *Am. J. Hum. Genet.* **44:** 388.

Structure of the Terminal Region of the Short Arm of Chromosome 16

Peter C. Harris and Douglas R. Higgs
MRC Molecular Haematology Unit
Institute of Molecular Medicine
John Radcliffe Hospital
Headington, Oxford OX3 9DU
United Kingdom

During the past few years, we have constructed a detailed physical map of the telomeric cytogenetic band of the short arm of chromosome 16, 16p13.3. This area, in common with the ends of most human chromosomes, is GC-rich and manifests distinctive structural and functional features. We have developed techniques for constructing a map of this area that will be helpful for studying the terminal regions of other chromosomes. The map has been useful for defining the positions of known genes and disease loci and has allowed detailed characterization of the telomere, revealing novel long-range polymorphism and chromosome rearrangements unique to chromosome ends.

This chapter describes:

❑ primary maps: hybrid map and genetic map

❑ strategies used to construct the long-range restriction map

❑ a detailed map of 16p13.3 up to and including the telomere

❑ long-range polymorphism of the 16p subtelomeric region

❑ examples of chromosome rearrangements specific to terminal regions

❑ genes and disease loci in 16p13.3

INTRODUCTION

It has become increasingly clear that the human genome is not a homogeneous structure but consists of a mosaic of DNA blocks, of similar GC content, juxtaposed to blocks of a very different base composition. These DNA blocks, called isochores, are on average greater than 300 kb in size. The characteristics of one isochore differ markedly from those of an isochore of a different GC content (for review, see Bernardi 1989). Characteristics that vary between isochores include the frequency of unmethylated CpG dinucleotides, gene density, the distribution of SINE and LINE repeats, and DNA replication timing (Bernardi et al. 1985). GC-poor and GC-rich regions largely correspond to G and R chromosome bands, respectively (for review, see Bickmore and Sumner 1989), bands that can be visualized with the use of base-specific DNA fluorochromes. A systematic study of the long-range structure of different genomic regions, each consisting of a different isochore, would be valuable to analyze further this variability in genomic composition.

Recent results on the location of the GC-richest isochores (H3), which constitute only 3% of the human genome, indicate that the majority of such isochores map to the telomeric regions of chromosomes (Saccone et al. 1992). The H3 isochores are especially interesting because they are particularly rich in CpG islands and are thought to contain a high concentration of genes (28% of human genes within 3% of total DNA; Mouchiroud et al. 1991). Clearly, a detailed study of such areas is likely to reveal a disproportionate amount of information about genes and disease loci.

Our interest has been in the terminal cytogenetic band of chromosome 16 (16p13.3) and has arisen from studies of the α-globin cluster, including a rare form of mental retardation associated with α-thalassemia (ATR-16) and the linkage of a locus for autosomal dominant polycystic kidney disease (ADPKD) to this region (Reeders et al. 1985). We have concentrated on constructing a physical map of this area to localize these disease loci. These studies have also allowed a detailed analysis of the structure of this interesting GC-rich area, which may provide insight into the functional significance of the isochore division of vertebrate genomes.

PRIMARY MAPS OF 16p13.3

In 1985, when an ADPKD locus was mapped to the short arm of chromosome 16, very few markers from this region had been described. Apart from the α-globin complex itself, only the uncloned red cell enzyme phosphoglycolate phosphatase (PGP) had been assigned to this interval (Povey et al. 1980; Reeders et al. 1986). Indeed, it was 2 more years before α-globin was precisely mapped to the band 16p13.3 by localizing it relative to the breakpoints of two chromosome rearrangements (Breuning et al. 1987a; Buckle et al. 1988). Subsequently, largely because of interest in ADPKD, many markers have been isolated from this region by several different groups (Table 1). These have been derived from chromosome-16-specific libraries produced either from flow-sorted chromosomes (Harris et al. 1987) or from somatic cell hybrids containing chromosome 16 as the only human material (Hyland et al. 1989; Keith et al. 1990; Breuning et al. 1990b). In addition, randomly isolated markers were assigned to this region (Xiao et al. 1987). In particular, minisatellite markers are enriched in this area (Julier et al. 1990; Armour et al. 1992; Royle et al. 1992), as in the terminal segments of other human chromosomes (Royle et al. 1988). Together these studies yielded sufficient markers to construct a preliminary map of 16p13.3. Two complementary maps were produced initially: a hybrid map and a genetic linkage map. The construction of these primary maps, which indicate the probe order, facilitated the subsequent production of a more detailed long-range restriction map.

Table 1 DNA probes within 16p13.3

Probe name	Locus symbol	Reference
pNFG400	D16S327	Wilkie et al. (1991)
α-Globin 5'HVR	D16S262	Jarman and Higgs (1988)
α-Globin 3'HVR	D16S85	Jarman et al. (1986)
CMM103	D16S252	Julier et al. (1990)
MS637	D16S307	Armour et al. (1992)
FR3-42	D16S21	Xiao et al. (1987)
EKMDA2	D16S83	Wolff et al. (1988)
MS205.2	D16S309	Royle et al. (1992)
NKIE5		Harris et al. (1990)
NKISP1	D16S146	Harris et al. (1990)
PNL56S	D16S145	Harris et al. (1990)
GGG1	D16S259	Germino et al. (1990)
CMM65	D16S84	Nakamura et al. (1988)
N54	D16S139	Himmelbauer et al. (1991)
SM7	D16S283	Harris et al. (1991)
26-6	D16S125	Breuning et al. (1990a)
VK5	D16S94	Hyland et al. (1990)
218EP6	D16S246	Snijdewint et al. (1990)
N2	D16S138	Himmelbauer et al. (1991)
CRI-090	D16S45	Keith et al. (1990)
CRI-327	D16S63	Keith et al. (1990)
24.1	D16S80	Breuning et al. (1987b)
LOM2B	D16S144	Harris et al. (1989)

The hybrid map

Interspecific somatic cell hybrids were produced that contain human derivative 16 chromosomes, deleted for various regions of 16p13.3. These chromosomes were isolated from individuals with a balanced or unbalanced product of a reciprocal translocation or a deleted chromosome, some of which are associated with the α-thalassemia/mental retardation syndrome (ATR-16) (see below). Chromosome 16 can be stably retained within an appropriate rodent cell line, after fusion, by selection for adenine phosphoribosyltransferase (APRT) (Deisseroth and Hendrick 1979; Zeitlin and Weatherall 1983), which is located on the distal long arm of chromosome 16. We have established a detailed somatic hybrid panel for the terminal region of chromosome 16p that complements the extensive panel of hybrids that now subdivides the whole of chromosome 16 (Callen et al. 1989; Chen et al. 1991).

Figure 1 shows the hybrid panel with breakpoints within 16p13.3 and the interval within which each probe lies. Probes from this region of chromosome 16 can be assigned to one of six intervals defined by the chromosomal breakpoints with the α-globin complex within the most telomeric (distal) interval.

The genetic linkage map

A genetic linkage map, in addition to defining the most likely order of probes, also provides a measure of distance between the markers, in terms of the recombination fraction. Two genetic maps covering all of chromosome 16 have been described (Julier et al. 1990; Keith et al. 1990). In addition, genetic maps of the terminal region of 16p13.3 have been produced by linkage analysis of probes within the reference CEPH pedigrees and in ADPKD families (Reeders et al. 1988; Germino et al. 1990). This linkage analysis indicated that the most likely order of probes was entirely consistent with that obtained by the hybrid map and localized the ADPKD locus on chromosome 16 to a region close to the probe CMM65 (Germino et al. 1990). Since the discovery of genetic heterogeneity of ADPKD (Kimberling et al. 1988; Romeo et al. 1988), the locus on chromosome 16 is referred to as the polycystic kidney disease 1 (PKD1) locus. Recent estimates indicate that ADPKD is due to mutation of the PKD1 locus in ~86% of families (Peters and Sandkuijl 1992).

One of the most interesting aspects of the genetic map is that there is much more male than female recombination across the whole terminal region (Germino et al. 1990; Julier et al. 1990) (discussed in detail below). The recombination frequency between the most distal marker analyzed, α-globin 3'HVR, and CMM65 is 0.07 in males but only 0.01 in females (Germino et al. 1990). Given the genomic average of 1% recombination for every 1 Mb of DNA, it was not clear whether the physical distance between these probes was closer to 1 Mb (expected from the

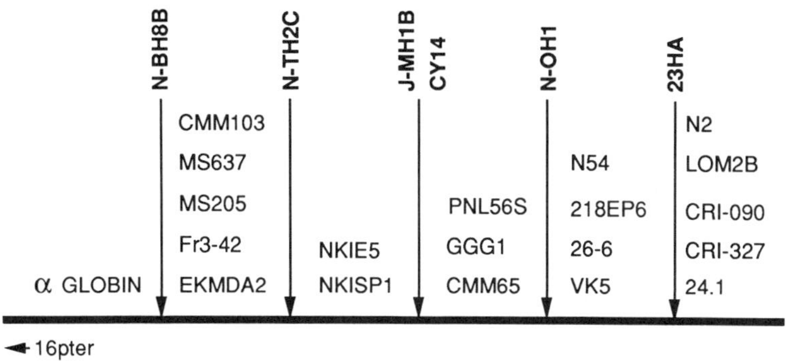

	N-BH8B	N-TH2C	J-MH1B CY14		N-OH1	23HA
	CMM103					N2
	MS637				N54	LOM2B
	MS205			PNL56S	218EP6	CRI-090
	Fr3-42	NKIE5		GGG1	26-6	CRI-327
α GLOBIN		EKMDA2	NKISP1	CMM65	VK5	24.1

◄ 16pter

Figure 1 A somatic cell hybrid map showing the location of six breakpoints subdividing 16p13.3. The interval to which 20 different probes from this region map is shown (Breuning et al. 1990b; Germino et al. 1990; Harris et al. 1990; Himmelbauer et al. 1991; K. Rack et al., in prep.).

level of female recombination) or 7 Mb (suggested by the male recombination rate). To determine the physical size of this region, a long-range restriction map was required.

STRATEGIES FOR THE CONSTRUCTION OF A LONG-RANGE RESTRICTION MAP

The restriction enzymes used to construct a long-range map digest rarely in genomic DNA because they recognize sites containing one or more CpG dinucleotides. The CpG dinucleotide is underrepresented in the genome (~20% of expected frequency), and most of those present (60–90%) are methylated and therefore resistant to cleavage. Many of the remaining unmethylated CpGs are clustered as islands and associated with genes (Bird 1986; Gardiner-Garden and Frommer 1987). The size of fragments produced by such enzymes is therefore dependent on the frequency of unmethylated CpGs within the region of study.

As previously noted, the frequency of CpG dinucleotides is not uniform throughout the genome. Preliminary long-range restriction mapping around the α-globin complex (Fischel-Ghodsian et al. 1987) revealed that this area is rich in CpG islands. Consequently, in the study of Fischel-Ghodsian et al. (1987), it was only possible to construct a map covering a few hundred kilobases around α-globin. Preliminary studies of the regions surrounding the more proximal markers, EKMDA2 and CMM65 (see Fig. 2A), also indicated a similar pattern of frequent digestion with CpG-recognizing restriction enzymes. This precluded physical linkage of these markers and indicated that different strategies would be required to complete a long-range map within this CpG-rich area.

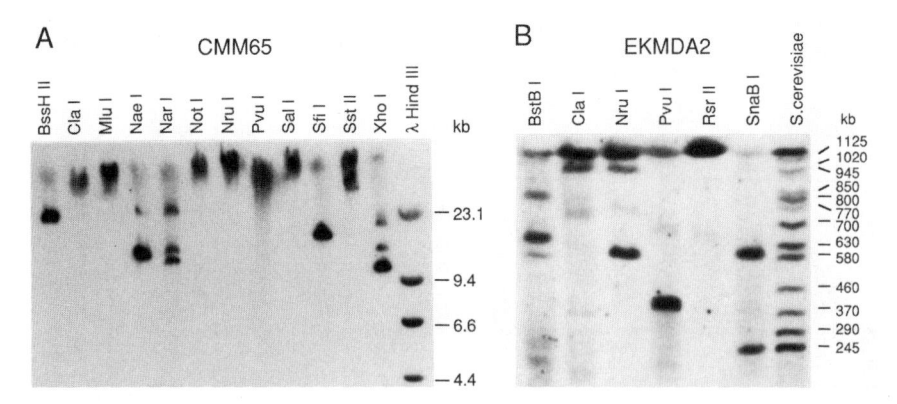

Figure 2 (A) Lymphoblast DNA digested with a variety of CpG-identifying enzymes, separated on a 0.5% conventional agarose gel and hybridized with the probe CMM65. Fragments smaller than ~30 kb are produced with many enzymes, especially those identifying sites consisting of only C and G nucleotides, indicating the frequency of these sites around this probe (see Fig. 3). Fragments too large to be resolved on this gel are seen as smears at the tops of the lanes. (B) Examples of the larger fragments produced with enzymes identifying sites containing A and T nucleotides as well as CpG (see Fig. 3). Digested lymphocyte DNA is resolved on a contour-clamped homogeneous electric field (CHEF) gel and hybridized with EKMDA2. S. cerevisiae chromosomes are used as a size standard.

One possibility was to isolate new markers from the 16p13.3 region. This was achieved for the interval between CMM65 and EKMDA2 by preparative field inversion gel electrophoresis of a 320-kb NotI fragment identified by CMM65 (Harris et al. 1990). One probe isolated in this way (PNL11bEH) mapped to the distal end of the NotI fragment, and using additional markers isolated at this new locus (NKISP1 and NKIE5), was mapped to a region 250 kb distal to CMM65. The NKIE5 locus proved useful in the construction of the long-range map, but it was clearly impractical to isolate additional markers for each small interval between probes, and so methods to produce larger restriction fragments for separation by pulsed field gel electrophoresis (PFGE) were explored.

The choice of restriction enzyme was found to have a marked effect on the size of fragments generated. Many CpG-identifying enzymes digest very frequently in this region, producing small fragments (<30 kb) (see Fig. 2A), whereas others digest less often, generating fragments suitable for linking more distant probes (Fig. 2B). If the enzymes are grouped by the base composition of their recognition sites, a correlation with the size of fragments identified by probes from 16p13.3 is seen (Fig. 3). Larger fragments are generated when recognition sites include the nucleotides A and T as well as at least one CpG.

Bird (1989) showed that enzymes recognizing GC-only sites including two CpGs (class a) usually digest within CpG islands, whereas en-

STRUCTURE OF 16p13.3   **115**

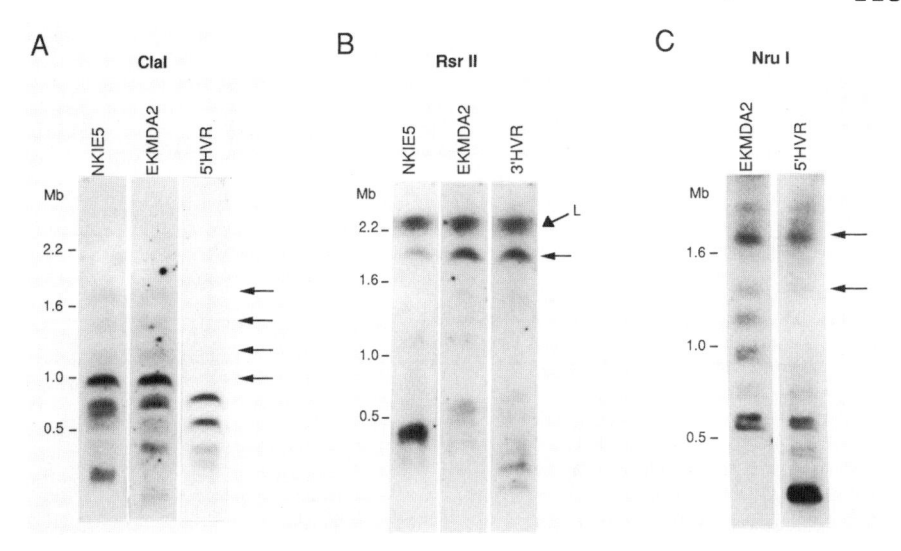

Figure 4 Examples of large fragments, resolved on CHEF gels, that link probes in 16p13.3. (A) A ClaI fragment of ~1 Mb and three larger partial digest fragments are detected by NKIE5 and EKMDA2, but not by α-globin 5'HVR. (B) A 1.85-Mb RsrII fragment is identified by all three probes. L indicates the zone of limiting mobility. (C) NruI fragments of 1.3 Mb and 1.7 Mb are detected by EKMDA2 and α-globin 5'HVR. (Reprinted, with permission, from Harris et al. 1990.)

than normal, indicating that the same 1.85-Mb RsrII fragment is shared by these probes and α-globin (Harris et al. 1990).

More precise long-range restriction mapping with complete and double digests was used to localize the markers. Subsequently, four additional probes (Fr3-42, CMM103, MS205.2, and MS637), which map by linkage and hybrid analysis to this region of the genome, have been positioned on the map (Harris et al. 1990; K. Rack et al., in prep.). From these results it was possible to produce the long-range restriction map shown in Figure 5. This map shows that the distance between α-globin and CMM65 is approximately 2 Mb, with EKMDA2 lying approximately midway between these two.

Localization of the 16p telomere

Figure 5 shows that the distal end of the large 1.85-Mb RsrII fragment is in a similar position to the ends of the most terminal ClaI and NruI fragments. These "sites" always digested readily, and small fragments that extend to these sites gave reproducibly broader bands when analyzed by PFGE, indicating variability in the size of fragments detected. The α-globin locus was the most distal marker studied on 16p, and so it seemed possible that the variable site just beyond α-globin might be the 16p telomere. We were able to test this hypothesis because a telomere-

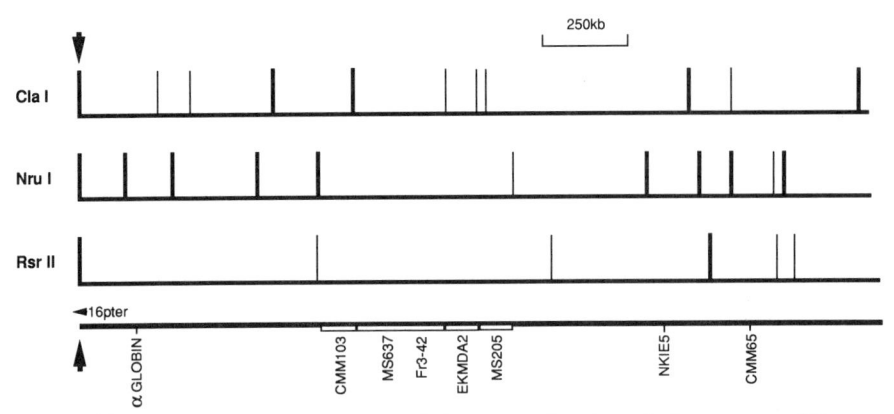

Figure 5 Long-range restriction map of the terminal region of chromosome 16, showing sites for the enzymes ClaI, NruI, and RsrII. Bold lines indicate sites that digest most readily in lymphocyte DNA. Some sites that partially digest and did not contribute to construction of the map are not shown. The arrows at the distal end of the map indicate coincident "sites" for all three enzymes, providing preliminary evidence that this is the position of the telomere.

associated probe, TelBam3.4, had been described which mapped to the telomeric region of 16p, as well as to several other subtelomeric regions, by in situ hybridization (Brown et al. 1990). Hybridization of α-globin 5'HVR and TelBam3.4 to DNA from somatic cell hybrids containing chromosome 16 showed that similar-sized fragments were identified with the enzymes MluI, NruI, and PvuI (Wilkie et al. 1991). Hybrid DNA was used to avoid detection of the other subtelomeric sites identified by TelBam3.4. Fragments identified by TelBam3.4 and 5'HVR were shown to be the same by analyzing DNA from a patient with a 62-kb deletion between 5'HVR and α-globin; the same novel fragments were identified by both probes (Wilkie et al. 1991). These results showed that α-globin lies only 170 kb from the 16p telomere and orients the α-globin cluster, such that the embryonic ζ-globin gene is closest to the telomere (tel-ζ2-α2-α1-cen). The localization of the α-globin cluster very close to the telomere provides an opportunity to compare the physical and genetic map in a region very close to the end of chromosome 16p.

Comparison of the genetic and physical maps of the terminal region of 16p

The genetic distances of the intervals from α-globin to EKMDA2 and from EKMDA2 to CMM65 have been estimated in two studies, both of which show a predominance of male recombination. Germino et al. (1990) found a value of 7.2 for the ratio of male to female recombination throughout the region from α-globin to the more proximal marker CRI-327. In this study, the male and female recombination frequencies (θ_m and θ_f) for the interval between α-globin and EKMDA2 were calculated to be $\theta_m = 0.03$ and $\theta_f = 0.005$, and between EKMDA2 and CMM65 as $\theta_m = 0.04$ and $\theta_f = 0.005$. Equivalent figures calculated by Julier et al. (1990) estimated three times more male recombination than female and a recombination fraction of $\theta_m = 0.042$ and $\theta_f = 0.015$ between α-globin and EKMDA2 and $\theta_m = 0.028$ and $\theta_f = 0.009$ from EKMDA2 to CMM65. If we compare these with the physical distances between the markers (~1.0 Mb between α-globin and EKMDA2 and ~0.9 Mb between EKMDA2 and CMM65), we can see that female recombination is slightly less frequent than would be expected at the genomic average of 1% recombination per megabase, whereas the male recombination is 3–4 times the average level. It is clear that this higher level of male recombination occurs in both of the intervals analyzed and is therefore not due to a single "hot spot" of male recombination.

A high male/female recombination ratio has been noted at the distal ends of many human chromosome linkage maps by Donis-Keller et al. (1987) and reviewed by Rouyer et al. (1990), whereas no sex difference is seen at the terminal ends of some genetic maps. In most cases, the linkage maps have not been physically tethered to the telomere, so it is not clear to what extent the recombination sex ratio detected at the end of the genetic map is reflecting the pattern close to the telomere. However, physical and genetic maps adjacent to some human telomeres have been constructed. At the tip of 4p, linkage analysis indicates a male/female recombination ratio of approximately 5 in the terminal 2 Mb of DNA, with male recombination approximately 4 times greater than the 1 Mb/1 cM genomic average (Buetow et al. 1991), closely reflecting the pattern seen at the end of the short arm of chromosome 16. In contrast, Burmeister et al. (1991) have compared the genetic and physical distance in an interval of 1.2 Mb, situated between 0.7 Mb and 1.9 Mb from the 21q telomere. In this case, a recombination rate 20 times higher than the 1 Mb/1 cM genomic average is seen in both males and females (Petersen et al. 1989). Preliminary studies of the telomere of 16q show that the existing genetic map (Keith et al. 1990) extends to within 230 kb of the chromosome end (Harris and Thomas 1992). On this map, the most distal interval between CRI089 and APRT has a recombination fraction of 0.03 in males and females (Keith et al. 1990). The physical size of this interval is not known, but it is greater than 1 Mb (P.C. Harris, unpubl.). Therefore, in this case recombination is not particularly enriched, and the recombination sex ratio is close to one.

Clearly, more chromosome ends need to be examined, but the emerging picture is one in which a higher than average rate of male recombination (in association with a low to average female level) is seen close to many telomeres. However, other patterns are also seen, such as a high general level of recombination or no enhanced level of recombination for either sex.

LENGTH POLYMORPHISM OF THE SUBTELOMERIC REGION OF 16p

Although initial studies of the end of chromosome 16p showed the telomere to be positioned 170 kb distal of α-globin, a more detailed analysis revealed that this structure (designated allele A) only accounts for approximately 70% of chromosomes, with three different subtelomeric structures accounting for the remainder (Wilkie et al. 1991). The detailed structure of the other variants (designated alleles B, C, and D) was determined by PFGE of individuals and cell lines with the different subtelomeric alleles using the enzymes *Mlu*I, *Nru*I, and *Pvu*I. Results obtained with *Mlu*I, which digests within the α-globin complex, showed that the size variation between alleles was due to polymorphism of the region distal to α-globin (Wilkie et al. 1991). Figure 6 shows DNA from hybrids that contain the A, D, B, or C allele digested with *Pvu*I and hybridized with α-globin 5'HVR. These results, supported by similar results with *Mlu*I and *Nru*I, indicate that the alleles D, B, and C are 75 kb, 180 kb, and 260 kb longer than the A allele, respectively. To test whether these fragments are indeed telomeric, hybrids containing each length allele were hybridized with telomere-associated sequences.

As described previously, the A allele contains the telomere-associated sequence TelBam3.4, and this probe also hybridizes to the terminal fragment of the C allele (Fig. 7A). However, the B allele has a different telomere-associated sequence, TelBam11 (Fig. 7A). This sequence, like TelBam3.4, hybridizes to many subtelomeric regions when analyzed by in situ hybridization (Brown et al. 1990). The mapping of TelBam11 to allele B was confirmed by in situ hybridization to metaphase spreads from a BB individual. Allele D does not hybridize with TelBam3.4 or TelBam11, but the consistent difference in size between the A and D allele seen with *Mlu*I-, *Nru*I-, and *Pvu*I-digested DNA suggests that the fragments detected with α-globin 5'HVR on D allele chromosomes are telomeric. The telomeric location of TelBam3.4 on the A and C alleles and TelBam11 on the B allele was confirmed by showing that these loci on chromosome 16p are sensitive to *Bal*-31 digestion.

The subtelomeric region of 16p exhibits long-range length polymorphism that places the α-globin locus 170 kb, 245 kb, 350 kb, or 430 kb from the telomere of the A, D, B, or C alleles, respectively (Fig. 7B). The proterminal region of chromosome 16p, consisting of different length alleles containing different telomere-associated sequences, suggests that this polymorphism has arisen by subtelomeric translocation between nonhomologous chromosomes. To learn the structure of the different alleles, we examined the divergence point between the A and B alleles.

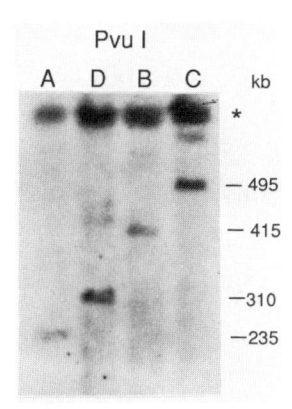

Figure 6 *Pvu*I-digested DNA of hybrids containing the subtelomeric length alleles, A (hybrid, 28/4), D (hybrid, J-PK4S), B (Hy145-19), or C (hybrid, J-BHIE) separated on a CHEF gel and hybridized with α-globin 5'HVR. Terminal *Pvu*I fragments of 235 kb, 310 kb, 415 kb, or 495 kb are seen for the A, D, B, and C alleles, respectively. The star indicates the zone of limiting mobility.

The divergence point between alleles A and B

The region distal to α-globin 5'HVR was analyzed by hybridization in an attempt to localize the position of divergence between the A and B alleles. One probe, pNFG400, positioned 135 kb telomeric to α-globin (Fig. 7B), identifies different chromosome-16-specific *Bgl*II fragments in hybrids containing different subtelomeric length alleles. This probe also hybridizes to four other non-16 fragments in genomic DNA. Comparison of the pNFG400/*Bgl*II alleles a, b, and c and the subtelomeric length alleles A, B, and C shows complete concordance between the two systems in 38 chromosomes analyzed. The frequency of the three alleles is 0.76, 0.24, and <0.01, for the A-a, B-b, and C-c alleles, respectively, in Caucasians; two D allele chromosomes analyzed both have the b, pNFG400/*Bgl*II allele. A possible explanation of the concordance between the polymorphic systems is that the pNFG400/*Bgl*II alleles cross the divergence point between the chromosome length alleles. Further analysis of the largest (a) allele showed this to be the case, with homology between the A and B alleles ceasing at a point approximately 10 kb distal to pNFG400. The structure at the position of divergence between

120 P.C. HARRIS AND D.R. HIGGS

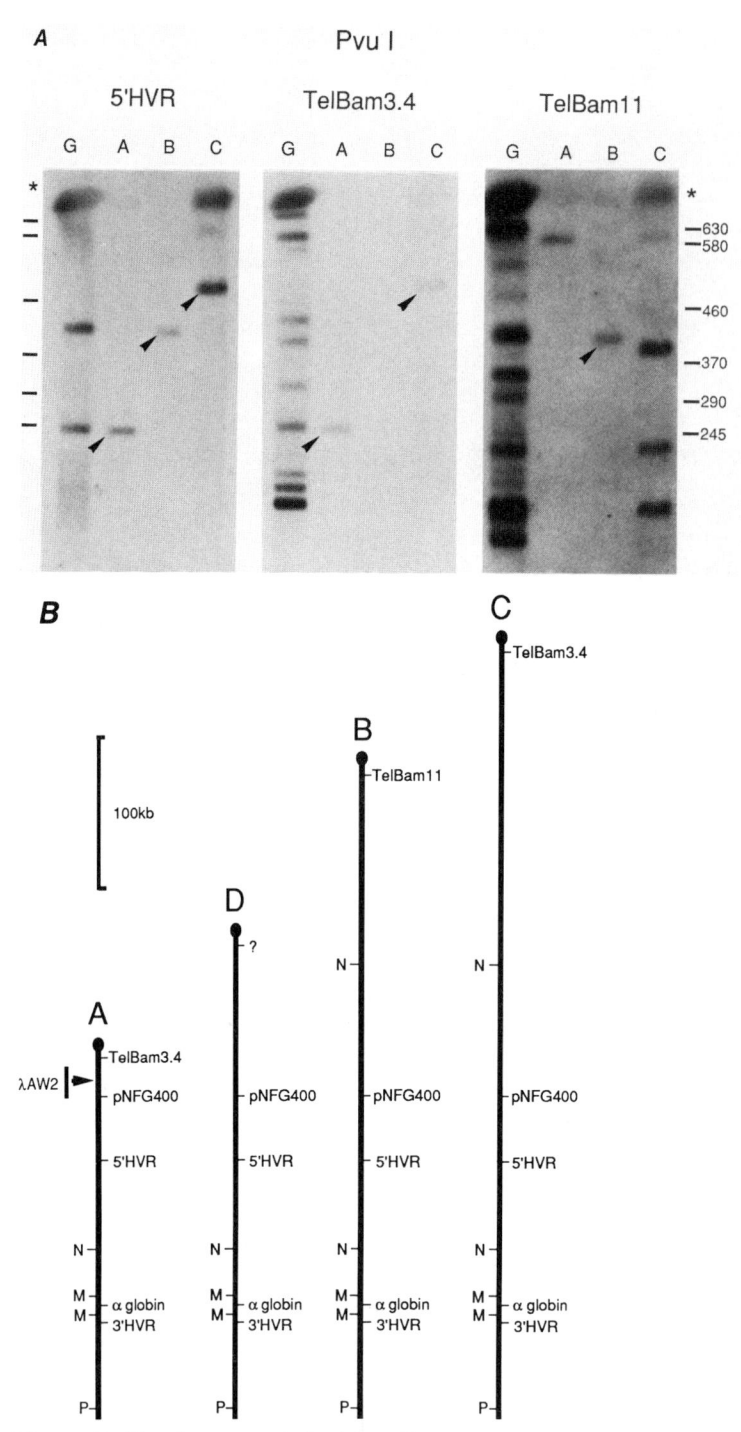

Figure 7 (*See facing page for legend.*)

the A and B length alleles is complex and consists of a region of ~4 kb where there is partial loss of homology between alleles, including a very large CA repeat (Wilkie and Higgs 1992) that differs in length with different alleles, before the final point of divergence is reached.

Analysis of the other loci identified by pNFG400 revealed that this probe hybridizes to sequences in the subtelomeric region of Xq and Yq, 9q, 10p, and 18p (Wilkie et al. 1991). The divergence point between the A and B length alleles therefore occurs within a region of subtelomeric repeats which are located at the ends of other, nonhomologous chromosomes. Analysis of allele a sequences beyond the divergence point indicate that these are homologous to the Xq and Yq sequences, whereas sequences associated with allele B are more closely related to the 9q, 10p, and 18p loci. The length alleles C and D are related to the B allele, beyond the divergence point, and probably result from a more distal exchange event or truncation involving the B allele.

Structural and functional consequences of the length polymorphism of 16pter

The structure of the proterminal region of chromosome 16 is complex, with chromosome 16 homologs often having nonhomologous subtelomeric regions: regions which have closer homology with the ends of other chromosomes. The location of the divergence point between the length alleles within a family of subtelomeric repeats and the presence of different telomere-associated sequences on different alleles suggest that the polymorphism has arisen by subtelomeric translocation with nonhomologous chromosomes. How frequent is this type of long-range polymorphism at the end of human chromosomes likely to be?

The terminal regions of several human chromosomes have now

Figure 7 CHEF gel of *Pvu*I-digested DNA of (lane G), a lymphoblastoid cell line from an AB individual; (lane A) allele A hybrid, P-SHHIM; (lane B) allele B hybrid, Hy145-19; (lane C) allele C hybrid J-BHIE, hybridized with α-globin 5′HVR (*left*), TelBam3.4 (*center*), and TelBam11 (*right*). The same fragments of 235 kb and 495 kb are detected by α-globin 5′HVR and TelBam3.4 (arrowed), but not TelBam11, in the allele A and allele C hybrids, respectively. The allele B fragment of 415 kb hybridizes with α-globin 5′HVR and TelBam11, but not TelBam3.4. The other bands detected with TelBam11 in the C allele hybrid are due to the presence of other chromosomes in this hybrid. Multiple fragments are seen in genomic DNA with TelBam3.4 and TelBam11 because they identify sites at the telomeres of many different chromosomes (Brown et al. 1990). The 600-kb fragment seen in the A and C hybrids with TelBam11 is the 16q telomere (see text). The star indicates the zone of limiting mobility. (Reprinted, with permission, from Wilkie et al. 1991; copyright by Cell Press.) (*B*) Long-range restriction map of the terminal regions of A, D, B, and C alleles of 16p. Allele D does not contain the telomere-associated sequences TelBam3.4 or TelBam11. The divergence point between the A allele and the other alleles is indicated by an arrow. λAW2 is a clone containing this divergence point (see below). Abbreviations: (M) *Mlu*I;(N) *Nru*I; (P) *Pvu*I.

been studied by PFGE: Xp/Yp (Brown 1988; Petit et al. 1988; Rappold and Lehrach 1988), Xq (Poustka et al. 1991), 4p (Doggett et al. 1989; Bates et al. 1990), and 21q (Burmeister et al. 1991); however, in none of these cases has length variation of the subtelomeric area been described. However, it is worth noting that most of these maps have been produced using DNA from a small number of different cell lines or individuals, and it is possible that variation would be seen if this analysis were extended to a larger number of subjects. The polymorphic distribution of the subtelomeric probe, TelBam3.4, in three different individuals analyzed by in situ hybridization (Brown et al. 1990), suggests that length polymorphism of the proterminal regions of other human chromosomes is likely. Recent studies in our laboratory of the structure of the 16q telomere indicate that this is indeed the case with long-range polymorphism of this subtelomeric region detected. The probe at the distal end of the 16q genetic map, CRI089 (Keith et al. 1990), has been linked to the telomere-associated sequence, TelBam11, on the majority of 16q telomeres. However, two other subtelomeric structures of 16q are seen, one of which contains TelBam3.4, and which result in length variation of up to 125 kb from that of the most frequent 16q allele (Harris and Thomas 1992). These findings indicate that length polymorphism of chromosome ends will probably be a feature common to other human chromosomes.

Cytological studies of chromosome pairing have suggested that pairing and formation of the synaptonemal complex are usually initiated close to the telomere (Rasmussen and Holm 1978; Bojko 1983; Chandley 1989), perhaps taking advantage of the tethering of telomeres to the nuclear matrix (de Lange 1992). The structures that we have described for the subtelomeric regions of chromosome 16 suggest that they may not be suitable areas in which to initiate pairing between homologs, if this is based on sequence homology; they are frequently nonhomologous and often have closer homology with other chromosome ends or even the other end of the same chromosome. It seems likely that pairing is initiated at sites proximal to these telomeric repeats, a hypothesis consistent with some cytological studies which suggest that to initiate pairing, telomeric recognition is not sufficient and more proximal regions of homology are required (Chandley 1989). The linkage disequilibrium seen between the telomere-associated sequence of each 16p length allele and the pNFG400/*Bgl*II allele indicates that recombination is suppressed in the nonhomologous region, suggesting that these sequences remain unpaired. The effect that these regions of nonhomology may have in promoting rearrangement between non-homologous chromosomes, such as cytogenetically invisible translocations (Lamb et al. 1989), is unclear. However, no increased frequency of translocations between chromosomes containing homologs of the pNFG400 family of subtelomeric repeats has been reported (Wilkie et al. 1991).

CHROMOSOME REARRANGEMENT OF 16p13.3

The proximity of the α-globin locus to the telomere of 16p, together with the simplicity by which deletions can be identified (producing the hematological phenotype of α-thalassemia; for review, see Higgs et al. 1989), provides a model system for studying the mechanism of deletion of terminal regions. Many small interstitial deletions involving α-globin and resulting in α-thalassemia have been described (Higgs et al. 1989). However, there are two other groups of rearrangements of this region: larger deletions (50–250 kb), which result in α-thalassemia alone, and deletion of a region of 1–2 Mb of 16p13.3 (Wilkie et al. 1990b), which results in α-thalassemia and mental retardation (ATR-16).

One case of α-thalassemia that is due to deletion of a region distal to α-globin, including the α-globin *cis*-acting regulatory sequence (HS-40) (Higgs et al. 1990), has been studied in detail (Wilkie et al. 1990a) and reveals a novel mechanism of human chromosome rearrangement. In this case, the chromosome has been truncated, and the resulting chromosome has been repaired by the direct addition of telomeric sequence (TTAGGG)$_n$ to the break. This chromosome is inherited through one generation and therefore shows that chromosome 16 can be stably inherited without telomere-associated repeats, indicating that telomeric sequence alone is sufficient to stabilize a human chromosome. In vitro analysis to test the ability of human telomerase to elongate from an oligonucleotide primer that matches the sequence at the break indicates that the minimal complementarity shown at the break with the RNA template of telomerase is sufficient to prime telomere synthesis (Morin 1992). Recent analysis has revealed that this mechanism of chromosome deletion may be more widely found, since we have identified three other truncations with breakpoints distal to α-globin (unpublished observations). In addition, a case of the ATR-16 syndrome L(BO) with a large truncation of 2 Mb of the end of chromosome 16 has been characterized and shown to be stabilized with telomeric sequence (Lamb et al. 1993). These results suggest that truncation, healed by the addition of telomeric sequence, may be a significant mechanism by which rearrangement occurs in the terminal regions of chromosomes and may account for cases of other contiguous gene syndromes associated with deletion of terminal chromosome regions (Gusella et al. 1985; Ledbetter et al. 1989).

Patients with the ATR-16 syndrome have a rather nonspecific clinical phenotype apart from α-thalassemia and mild to moderate mental handicap (Wilkie et al. 1990b). Part of this heterogeneity may be due to the size of the region of 16p that is deleted (summarized in Fig. 8) but also is probably due to the fact that many cases are aneuploid for a second chromosome region. Two ATR-16 cases have inherited an unbalanced product of a reciprocal translocation. One of these cases, OD, has a chromosome rearrangement that was not detected by cytogenetic analysis and was only determined by molecular analysis and in situ hybridization (Lamb et al. 1989). This type of submicroscopic terminal translocation is another form of rearrangement specific for terminal chromosome regions, and inheritance of unbalanced products from such exchanges may account for a significant proportion of unexplained mental handicap and dysmorphism.

Three other ATR-16 cases are due to de novo translocations resulting in an unbalanced karyotype with the derivative 16 chromosome deleted for a region of 16p (see Fig. 8) and trisomic for an additional chromosomal region. In two of these cases, the region of trisomy was not identifiable by cytogenetic analysis, and we have employed the method of "reverse chromosome painting" (Carter et al. 1992) to determine their origin. This technique involves making a "paint" from the flow-sorted abnormal chromosome and applying this paint to normal metaphase spreads. In both cases, the extra material was identified as the distal two-thirds of chromosome 9p (K. Rack et al., in prep.). Three other ATR-16 cases that are monosomic for 1–2 Mb of 16p13.3 appear normal by conventional cytogenetic analysis, and therefore the basis of the rearrangement (truncation, deletion, or translocation) is not clear. To determine more precisely a critical area of 16p13.3 that is associated with mental retardation, more patients with pure monosomy of this region, such as the case due to truncation (Lamb et al. 1993), will need to be analyzed.

DISEASE LOCI AND GENE FREQUENCY IN 16p13.3

One major reason for the detailed study of this region has been the interest in PKD1. Since results indicating that the PKD1 locus is close to the probe CMM65 (Germino et al. 1990), which has been mapped to a region 2 Mb from the 16p telomere (Harris et al. 1990), considerable progress to map the PKD1 region genetically and physically has been made. Several new markers that map close to PKD1 have been described (Breuning et al. 1990b; Hyland et al. 1990; Himmelbauer et al. 1991), including a CA repeat polymorphism on the proximal side of the locus (Harris et al. 1991). Recently, analysis of a number of PKD1 families with crossovers close to the disease locus have allowed the refinement of the candidate region to an area between the markers GGG1 and 26.6PROX (Somlo et al. 1992). Long-range restriction mapping of this interval shows that PKD1 lies within an area of approximately 750 kb (Germino et al. 1992), which can be refined to an area of approximately 620 kb if results obtained with SM7 (Harris et al. 1991) are included (Fig. 8). This area is very rich in CpG islands and has now largely been cloned as a cosmid contig (Germino et al. 1992) with more than 20 sets of cDNA clones identified within the candidate region. A description of

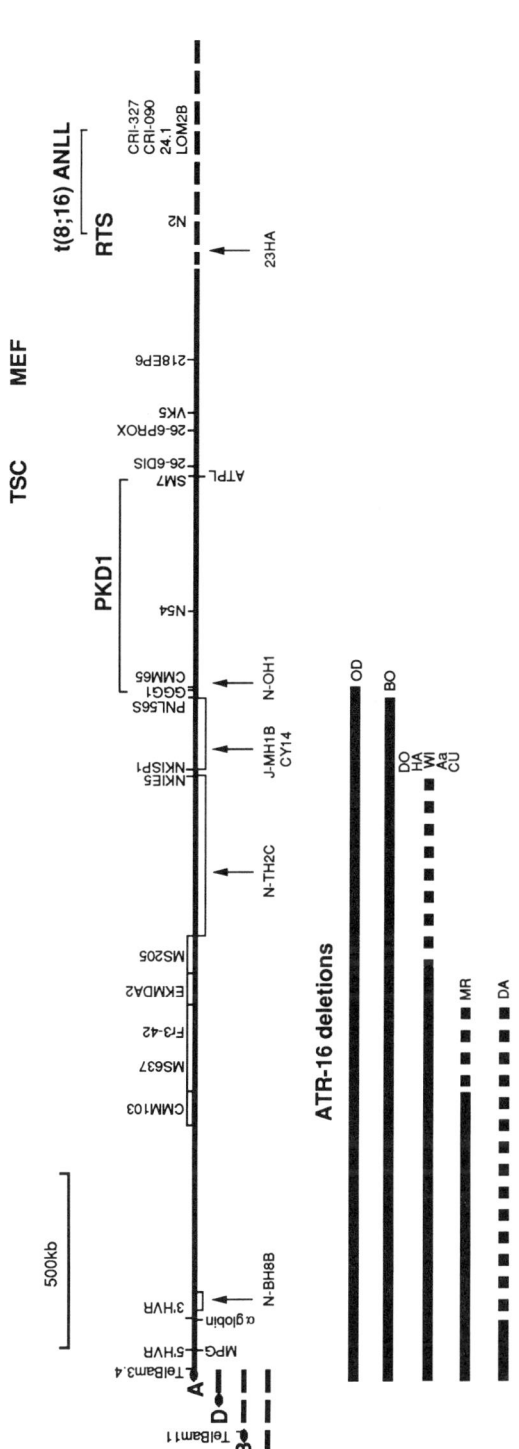

Figure 8 A summary map of the terminal three megabases of the short arm of chromosome 16. The locations of probes are shown above the line and the positions of cloned genes (Gillespie et al. 1991; M.A. Vickers et al., in prep.) and hybrid breakpoints below the line. The distances between markers in the proximal region, indicated by dashed lines, are unknown. The polymorphic subtelomeric regions A, D, B, and C are shown at the distal end of the map. Regions containing the disease loci PKD1 (Harris et al. 1991; Germino et al. 1992) and RTS (Breuning et al. 1993) and the breakpoint of the t(8;16) associated with ANLL (Wessels et al. 1991) are indicated. The preliminary locations of disease loci mapped to this region by linkage analysis, MEF (Pras et al. 1992) and TSC (Kandt et al. 1992), are also shown. The size of regions of monosomy in cases of the ATR-16 syndrome are illustrated as solid bars below the map. Dashed lines indicate the region within which the breakpoints fall. (Summarized from Harris et al. 1990, 1991; Wilkie et al. 1990b, 1991; Germino et al. 1992; K. Rack et al., in prep. and unpubl.)

linkage disequilibrium between the proximal flanking probe, VK5, and PKD1 may provide a clue to the most likely location of the PKD1 gene (Pound et al. 1992).

In addition to the previously identified disease loci mapped to 16p13.3—PKD1, α-globin, and the determinants of normal development deleted in ATR-16 patients—several other diseases have recently been mapped to this area. Genetic linkage analysis of families with the recessive disorder, familial Mediterranean fever (MEF), have shown linkage with CMM65; a maximal LOD score of 19.85 was obtained at a point (10 cM in male meiosis) proximal to this marker (Pras et al. 1992).

Two unrelated chromosome translocations in patients with Rubinstein-Taybi syndrome (RTS) were both found to have breakpoints in 16p13.3 and so provided the clue that RTS maps to this area (Breuning et al. 1993). The breakpoints of both translocations have been mapped to a single cosmid that is located close to the probe N2 (Fig. 8). The discovery of submicroscopic deletions of this region in 6 of 24 other RTS patients confirms this as the site of the RTS locus and should help with cloning the gene.

The most recent disease locus to be mapped by genetic linkage analysis to the terminal band of the chromosome 16 short arm is a major locus for tuberous sclerosis (TSC) (Kandt et al. 1992). In this study, five TSC families which are not linked to other TSC loci showed linkage with the marker SM7 with a cumulative LOD score of 9.52 at $\hat{\theta}$ = 0.02. These results indicate that there is a TSC locus near the PKD1 region (Fig. 8). The breakpoint of one cancer-related rearrangement, t(8;16)(p11;p13), associated with acute nonlymphocytic leukemia (ANLL), has been localized to the region between the probes CRI-327 and the RTS region (Fig. 8) (Wessels et al. 1991; Breuning et al. 1993).

The discovery of many disease loci in this small region seems to support the argument that such regions, rich in GC nucleotides and CpG islands and situated in terminal regions, are unusually gene-rich (Saccone et al. 1992). It is now possible to test this by systematic searches of 16p13.3 for gene loci. Vyas et al. (1992) have analyzed the terminal 165 kb of 16p, proximal to the subtelomeric repeat sequences for genes. In addition to the embryonic (ζ2), the fetal/adult (α2,α1) globin genes, and the θ1 gene, four widely expressed genes were identified. One of these encodes the DNA repair enzyme methyl adenine DNA glycosylase (MPG; M.A. Vickers et al., in prep.). If a total number of human genes of between 5×10^4 and 1×10^5 is assumed and the total size of the genome is estimated to be 3.3×10^9 bp, then the gene frequency in this terminal interval is 1.5 –3 times that expected if genes were evenly distributed.

A second region within 16p13.3 that has been studied in detail is that containing the PKD1 locus (Germino et al. 1992). In this area, 20 cDNAs have been isolated in 500 kb, suggesting a similar gene density to that found adjacent to the telomere. This gene density is approximately 3 times higher than that detected in one well-studied AT-rich region of the genome containing the cystic fibrosis gene (Rommens et al. 1989). It has been estimated that the GC-richest fraction of DNA (H3 isochore), which represents 3% of total human DNA, contains 28% of genes (Mouchiroud et al. 1991), indicating a density of one gene every 3.5–7 kb. This level is three to five times higher than that observed in 16p13.3, although this region does appear to fit the H3 criteria for both GC content and frequency of CpG islands. It may be that the extent of enrichment for genes in the H3 fraction has been overestimated, or it may be that other genes will be found in the areas searched, perhaps not associated with CpG islands. Another factor influencing the calculations is the figure assumed for the total number of genes; this may be an overestimate with alternate splicing accounting for more protein diversity. Preliminary analysis of the types of genes in this terminal area shows that a mixture of constitutively expressed and tissue-specific genes are found in close proximity (Vyas et al. 1992).

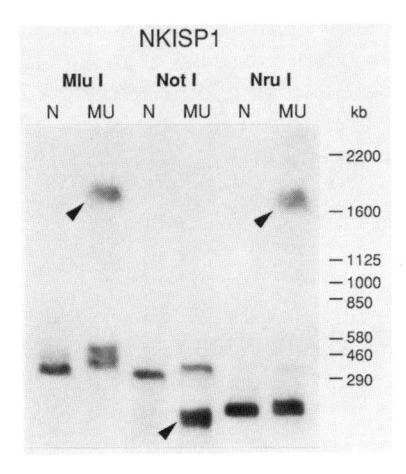

Figure 9 Detection of the chromosome breakpoint in individual MU who has the karyotype 46XY; t(5;16)(q31.1-3;p13.3). The hybrid J-MHIB contains the MU der16 chromosome which has a breakpoint between the probes NKISP1 and PNL56S (see Fig. 8). Breakpoint fragments (arrowed) are seen with *Mlu*I (~1.8 Mb), *Not*I (~100 kb), and *Nru*I (~1.7 Mb) with digested MU DNA (N is a normal control) separated on a CHEF gel and hybridized with NKISP1. The small fragment found with *Not*I-digested DNA indicates that the breakpoint must lie within 100 kb of NKISP1.

CONCLUSIONS

We have constructed a detailed physical map of the terminal region of 16p13.3 (Fig. 8). This map ensures that new probes can be rapidly sublocalized using the hybrid panel and PFGE. Likewise, the breakpoint of any chromosome rearrangement can be quickly localized by hybridizing mapped probes to the abnormal chromosome, either directed by in situ hybridization or after isolation in a somatic cell hybrid or by chromosome sorting (K. Rack et al., in prep.). The precise location of the breakpoint can then be determined by PFGE (Fig. 9).

Detailed analysis of the structure of this region has identified a novel long-range polymorphism of the subtelomeric region. The entire area of 3 Mb is GC-rich and contains a high frequency of unmethylated CpG dinucleotides. A higher than average density of genes and a relatively large number of disease loci have now been mapped to this area. It is clear that terminal regions such as this are susceptible to novel forms of chromosome rearrangement, which may be a significant cause of human genetic disease, including mental handicap and dysmorphism. It seems likely that emphasis on the genomic analysis of telomeric regions, besides shedding light on the special structures found in these areas, is likely to yield a wealth of information on genes and disease loci.

Acknowledgments

We thank all the individuals involved with this project whose work we have summarized here. We thank Caroline Bastable and Liz Rose for typing and Sir David Weatherall for support. Martijn Breuning is thanked for a preprint of the RT1 manuscript. This work was funded by the Medical Research Council, the Wellcome Trust, and Action Research for the Crippled Child.

References

Antequera, F., J. Boyes, and A. Bird. 1990. High levels of de novo methylation and altered chromatin structure at CpG islands in cell lines. *Cell* **62:** 503.

Armour, J.A.L., G. Vergnaud, M. Crosier, and A.J. Jeffreys. 1992. Isolation of human minisatellite loci detected by synthetic tandem repeat probes: Direct comparison with cloned DNA fingerprinting probes. *Hum. Mol. Genet.* **1:** 319.

Bates, G.P., M.E. MacDonald, S. Baxendale, Z. Sedlacek, S. Youngman, D. Romano, W.L. Whaley, B.A. Allitto, A. Poustka, J.F. Gusella, and H. Lehrach. 1990. A yeast artificial chromosome telomere clone spanning a possible location of the Huntington disease gene. *Am. J. Hum. Genet.* **46:** 762.

Bernardi, G. 1989. The isochore organization of the human genome. *Annu. Rev. Genet.* **23**: 637.

Bernardi, G., B. Olofsson, J. Filipski, M. Zerial, J. Salinas, G. Cuny, M. Meunier-Rotival, and F. Rodier. 1985. The mosaic genome of warm-blooded vertebrates. *Science* **228**: 953.

Bickmore, W.A. and A.T. Sumner. 1989. Mammalian chromosome banding—An expression of genome organization. *Trends Genet.* **5**: 144.

Bird, A.P. 1986. CpG-rich islands and the function of DNA methylation. *Nature* **321**: 209.

―――. 1989. Two classes of observed frequency for rare-cutter sites in CpG islands. *Nucleic Acids Res.* **17**: 9485.

Bojko, M. 1983. Human meiosis VIII. Chromosome pairing and formation of the synaptonemal complex in oocytes. *Carlsberg Res. Comm.* **48**: 457.

Breuning, M.H., K. Madan, M. Verjaal, J.T. Wijnen, P. Meera Khan, and P.L. Pearson. 1987a. Human α-globin maps to pter-p13.3 in chromosome 16 distal to PGP. *Hum. Genet.* **76**: 287.

Breuning, M.H., S.T. Reeders, H. Brunner, J.W. Ijdo, J.J. Saris, A. Verwest, G.J.B. van Ommen, and P.L. Pearson. 1987b. Improved early diagnosis of adult polycystic kidney disease with flanking DNA markers. *Lancet* **II**: 1359.

Breuning, M.H., F.G.M. Snijdewint, J.R. Smits, J.G. Dauwerse, J.J. Saris, and G.J.B. van Ommen. 1990a. A TaqI polymorphism identified by 26-6 (D16S125) proximal to the locus affecting adult polycystic kidney disease (PKD1) on chromosome 16. *Nucleic Acids Res.* **18**: 3106.

Breuning, M.H., H.G. Dauwerse, G. Fugazza, J.J. Saris, L. Spruit, H. Wijnen, N. Tommerup, C.B. van der Hagen, K. Imaizumi, Y. Kuroki, M.-J. van den Boogaard, J.M. de Pater, E.C.M. Mariman, B.C.J. Hamel, H. Himmelbauer, A.-M. Frischauf, R.L. Stallings, G.C. Beverstock, G.-J.B. van Ommen, and R.C.M. Hennekam. 1993. Rubinstein-Taybi syndrome caused by submicroscopic deletions within 16p13.3. *Am. J. Hum. Genet.* (in press).

Breuning, M.H., F.G.M. Snijdewint, H. Brunner, A. Verwest, J.W. Ijdo, J.J. Saris, J.G. Dauwerse, L. Blonden, T. Keith, D.F. Callen, V.J. Hyland, G.H. Xiao, G. Scherer, D.R. Higgs, P. Harris, L. Bachner, S.T. Reeders, G. Germino, P.L. Pearson, and G.J.B. van Ommen. 1990b. Map of 16 polymorphic loci on the short arm of chromosome 16 close to the polycystic kidney disease gene (PKD1). *J. Med. Genet.* **27**: 603.

Brown, W.R.A. 1988. A physical map of the human pseudoautosomal region. *EMBO J.* **7**: 2377.

Brown, W.R.A., P.J. MacKinnon, A. Villasanté, N. Spurr, V.J. Buckle, and M.J. Dobson. 1990. Structure and polymorphism of human telomere-associated DNA. *Cell* **63**: 119.

Buckle, V.J., D.R. Higgs, A.O.M. Wilkie, M. Super, and D.J. Weatherall. 1988. Localisation of human α globin to 16p13.3→pter. *J. Med. Genet.* **25**: 847.

Buetow, K.H., R. Shiang, P. Yang, Y. Nakamura, G.M. Lathrop, R. White, J.J. Wasmuth, S. Wood, L.D. Berdahl, N.J. Leysens, T.M. Ritty, M.E. Wise, and J.C. Murray. 1991. A detailed multipoint map of human chromosome 4 provides evidence for linkage heterogeneity and position-specific recombination rates. *Am. J. Hum. Genet.* **48**: 911.

Burmeister, M., S. Kim, E.R. Price, T. de Lange, U. Tantravahi, R.M. Myers, and D.R. Cox. 1991. A map of the distal region of the long arm of human

chromosome 21 constructed by radiation hybrid mapping and pulsed-field gel electrophoresis. *Genomics* **9**: 19.
Callen, D.F., V.J. Hyland, E.G. Baker, A. Fratini, A.K. Gedeon, J.C. Mulley, K.E.W. Fernandez, M.H. Breuning, and G.R. Sutherland. 1989. Mapping the short arm of human chromosome 16. *Genomics* **4**: 348.
Carter, N.P., M.A. Ferguson-Smith, M.T. Perryman, H. Telenius, A.H. Pelmear, M.A. Leversha, M.T. Glancy, S.L. Wood, K. Cook, H.M. Dyson, M.E. Ferguson-Smith, and L.R. Willatt. 1992. Reverse chromosome painting: A method for the rapid analysis of aberrant chromosomes in clinical cytogenetics. *J. Med. Genet.* **29**: 299.
Chandley, A.C. 1989. Asymmetry in chromosome pairing: A major factor in de novo mutation and the production of genetic disease in man. *J. Med. Genet.* **26**: 546.
Chen, L.Z., P.C. Harris, S. Apostolou, E. Baker, K. Holman, S.A. Lane, J.K. Nancarrow, S.A. Whitmore, R.L. Stallings, C.E. Hilderbrand, R.I. Richards, G.R. Sutherland, and D.F. Callen. 1991. A refined physical map of the long arm of human chromosome 16. *Genomics* **10**: 308.
de Lange, T. 1992. Human telomeres are attached to the nuclear matrix. *EMBO J.* **11**: 717.
Deisseroth, A. and D. Hendrick. 1979. Activation of phenotypic expression of human globin genes from non-erythroid cells by chromosome-dependent transfer to tetraploid mouse erythroleukemia cells. *Proc. Natl. Acad. Sci.* **76**: 2185.
Doggett, N.A., J.-F. Cheng, C.L. Smith, and C.R. Cantor. 1989. The Huntington disease locus is most likely within 325 kilobases of the chromosome 4p telomere. *Proc. Natl. Acad. Sci.* **86**: 10011.
Donis-Keller, H., P. Green, C. Helms, S. Cartinhour, B. Weiffenback, K. Stephens, T.P. Keith, D.W. Bowden, D.R. Smith, E.S. Lander, D. Botstein, G. Akots, K.S. Rediker, T. Gravius, V.A. Brown, M.B. Rising, C. Parker, J.A. Powers, D.E. Watt, E.R. Kauffman, A. Bricker, P. Phipps, H. Muller-Kahle, R.R. Fulton, S. Ng, J.W. Schumm, J.C. Braman, R.G. Knowlton, D.F. Barker, S.M. Crooks, S.E. Lincoln, M.J. Daly, and J. Abrahamson. 1987. A genetic linkage map of the human genome. *Cell* **51**: 319.
Fischel-Ghodsian, N., R.D. Nicholls, and D.R. Higgs. 1987. Long range genome structure around the human α-globin complex analysed by PFGE. *Nucleic Acids Res.* **15**: 6197.
Gardiner-Garden, M. and M. Frommer. 1987. CpG islands in vertebrate genomes. *J. Mol. Biol.* **196**: 261.
Germino, G.G., D. Weinstat-Saslow, H. Himmelbauer, G.A.J. Gillespie, S. Somlo, B. Wirth, N. Barton, K.L. Harris, A.-M. Frischauf, and S.T. Reeders. 1992. The gene for autosomal dominant polycystic kidney disease lies in a 750-kb CpG-rich region. *Genomics* **13**: 144.
Germino, G.G., N.J. Barton, J. Lamb, D.R. Higgs, P. Harris, G.H. Xiao, G. Scherer, Y. Nakamura, and S.T. Reeders. 1990. Identification of a locus which shows no genetic recombination with the autosomal dominant polycystic kidney disease gene on chromosome 16. *Am. J. Hum. Genet.* **46**: 925.
Gillespie, G.A.J., S. Somlo, G.G. Germino, D. Weinstat-Saslow, and S.T. Reeders. 1991. CpG island in the region of an autosomal dominant polycystic kidney disease locus defines the 5' end of a gene encoding a putative proton

channel. *Proc. Natl. Acad. Sci.* **88**: 4289.

Gusella, J.F., R.E. Tanzi, P.I. Bader, M.C. Phelan, R. Stevenson, M.R. Hayden, K.J. Hofman, A.G. Faryniarz, and K. Gibbons. 1985. Deletion of Huntingdon's disease-linked G8(D4S10) locus in Wolf-Hirschhorn syndrome. *Nature* **318**: 75.

Harris, P. and S. Thomas. 1992. Length polymorphism of the subtelomeric regions of both the short (p) and long arms (q) of chromosome 16. *Cytogenet. Cell Genet.* **60**: 171.

Harris, P.C., S.T. Reeders, O.J. Lehmann, and R.E. Tanzi. 1989. Linkage analysis of DNA probes from 16p. *Cytogenet. Cell Genet.* **51**: 1011.

Harris, P.C., N.J. Barton, D.R. Higgs, S.T. Reeders, and A.O.M. Wilkie. 1990. A long-range restriction map between the α-globin complex and a marker closely linked to the polycystic kidney disease (PKD1) locus. *Genomics* **7**: 195.

Harris, P., M. Lalande, H. Stroh, G. Bruns, A. Flint, and S.A. Latt. 1987. Construction of a chromosome 16-enriched phage library and characterization of several DNA segments from 16p. *Hum. Genet.* **77**: 95.

Harris, P.C., S. Thomas, P.J. Ratcliffe, M.H. Breuning, E. Coto, and C. Lopez-Larrea. 1991. Rapid genetic analysis of families with polycystic kidney disease by means of a microsatellite marker. *Lancet* **338**: 1484.

Higgs, D.R., M.A. Vickers, A.O.M. Wilkie, I.-M. Pretorius, A.P. Jarman, and D.J. Weatherall. 1989. A review of the molecular genetics of the human α-globin gene cluster. *Blood* **73**: 1081.

Higgs, D.R., W.G. Wood, A.P. Jarman, J. Sharpe, J. Lida, I.-M. Pretorius, and H. Ayyub. 1990. A major positive regulatory region located far upstream of the human α-globin gene locus. *Genes Dev.* **4**: 1588.

Himmelbauer, H., G.G. Germino, I. Ceccherini, G. Romeo, S.T. Reeders, and A.-M. Frischauf. 1991. Saturating the region of the polycystic kidney disease gene with *Not*I linking clones. *Am. J. Hum. Genet.* **48**: 325.

Hyland, V.J., K.E.W. Fernandez, D.F. Callen, R.N. MacKinnon, E. Baker, K. Friend, and G.R. Sutherland. 1989. Assignment of anonymous DNA probes to specific intervals of human chromosomes 16 and X. *Hum. Genet.* **83**: 61.

Hyland, V.J., G.K. Suthers, K. Friend, R.N. MacKinnon, D.F. Callen, M.H. Breuning, T. Keith, V.A. Brown, P. Phipps, and G.R. Sutherland. 1990. Probe, VK5B, is located in the same interval as the autosomal dominant adult polycystic kidney disease locus, PKD1. *Hum. Genet.* **84**: 286.

Jarman, A.P. and D.R. Higgs. 1988. A new hypervariable marker for the human α-globin gene cluster. *Am. J. Hum. Genet.* **43**: 249.

Jarman, A.P., R.D. Nicholls, D.J. Weatherall, J.B. Clegg, and D.R. Higgs. 1986. Molecular characterisation of a hypervariable region downstream of the human α-globin gene cluster. *EMBO J.* **5**: 1857.

Julier, C., Y. Nakamura, M. Lathrop, P. O'Connell, M. Leppert, T. Mohandas, J.-M. Lalouel, and R. White. 1990. A primary map of 24 loci on human chromosome 16. *Genomics* **6**: 419.

Kandt, R.S., J.L. Haines, M. Smith, H. Northrup, R.J.M. Gardner, M.P. Short, K. Dumars, E.S. Roach, S. Steingold, S. Wall, S.H. Blanton, P. Flodman, D.J. Kwiatkowski, A. Jewell, J.L. Weber, A.D. Roses, and M.A. Pericak-Vance. 1992. Linkage of an important gene locus for tuberous sclerosis to a chromosome 16 marker for polycystic kidney disease. *Nature Genet.* **2**: 37.

Keith, T.P., P. Green, S.T. Reeders, V.A. Brown, P. Phipps, A. Bricker, K. Falls,

K.S. Rediker, J.A. Powers, C. Hogan, C. Nelson, R. Knowlton, and H. Donis-Keller. 1990. Genetic linkage map of 46 DNA markers on human chromosome 16. *Proc. Natl. Acad. Sci.* **87:** 5754.

Kimberling, W.J., P.R. Fain, J.B. Kenyon, D. Goldgar, E. Sujansky, and P.A. Gabow. 1988. Linkage heterogeneity of autosomal dominant polycystic kidney disease. *N. Engl. J. Med.* **319:** 913.

Lamb, J., A.O.M. Wilkie, P.C. Harris, V.J. Buckle, R.H. Lindenbaum, N.J. Barton, S.T. Reeders, D.J. Weatherall, and D.R. Higgs. 1989. Detection of breakpoints in submicroscopic chromosomal translocation, illustrating an important mechanism for genetic disease. *Lancet* **2:** 819.

Lamb, J., P.C. Harris, A.O.M. Wilkie, W.G. Wood, S.G. Dauwerse, and D.R. Higgs. 1993. De novo truncation of chromosome 16p and healing with $(TTAGG)_n$ in the α thalassaemia/mental retardation syndrome (ATR-16). *Am J. Hum. Genet.* (in press).

Ledbetter, D.H., S.A. Ledbetter, P. vanTuinen, K.M. Summers, T.J. Robinson, Y. Nakamura, R. Wolff, R. White, D.F. Barker, M.R. Wallace, F.S. Collins, and W.B. Dobyns. 1989. Molecular dissection of a contiguous gene syndrome: Frequent submicroscopic deletions, evolutionarily conserved sequences, and a hypomethylated "island" in the Miller-Dieker chromosome region. *Proc. Natl. Acad. Sci.* **86:** 5136.

Morin, G.B. 1992. Recognition of a chromosome truncation site associated with α-thalassaemia by human telomerase. *Nature* **353:** 454.

Mouchiroud, D., G. d'Onofrio, B. Aïssani, G. Macaya, C. Gautier, and G. Bernardi. 1991. The distribution of genes in the human genome. *Gene* **100:** 181.

Nakamura, Y., C. Martin, K. Krapcho, P. O'Connell, M. Leppert, G.M. Lathrop, J.-M. Lalouel, and R. White. 1988. Isolation and mapping of a polymorphic DNA sequence (pCMM65) on chromosome 16 [D16S84]. *Nucleic Acids Res.* **16:** 3122.

Peters, D.J.M. and L.A. Sandkuijl. 1992. Genetic heterogeneity of polycystic kidney disease in Europe. In *Polycystic kidney disease: Contributions to nephrology; 97* (ed. M.H Breuning et al.), p. 128. Karger, Basel.

Petersen, M.B., S.A. Slaugenhaupt, J.G. Lewis, A.C. Warren, A. Chakravarti, and S.E. Antonarakis. 1989. A genetic linkage map of 24 loci on human chromosome 21. *Am. J. Hum. Genet.* **45:** A157.

Petit, C., J. Levilliers, and J. Weissenbach. 1988. Physical mapping of the human pseudo-autosomal region; comparison with genetic linkage map. *EMBO J.* **7:** 2369.

Pound, S.E., A.D. Carothers, P.M. Pignatelli, A.M. Macnicol, M.L. Watson, and A.F. Wright. 1992. Evidence for linkage disequilibrium between *D16S94* and the adult onset polycystic kidney disease (PKD1) gene. *J. Med. Genet.* **29:** 247.

Poustka, A., A. Dietrich, G. Langenstein, D. Toniolo, S.T. Warren, and H. Lehrach. 1991. Physical map of human Xq27-qter: Localizing the region of the fragile X mutation. *Proc. Natl. Acad. Sci.* **88:** 8302.

Povey, S., S.J. Jeremiah, R.F. Barker, D.A. Hopkinson, E.B. Robson, and P.J.L. Cook. 1980. Assignment of the human locus determining phosphoglycolate phosphatase (PGP) to chromosome 16. *Ann. Hum. Genet..* **43:** 241.

Pras, E., I. Aksentijevich, L. Gruberg, J.E. Balow, L. Prosen, M. Dean, A.D. Steinberg, M. Pras, and D.L. Kastner. 1992. Mapping of a gene causing familial

Mediterranean fever to the short arm of chromosome 16. *N. Engl. J. Med.* **326**: 1509.

Rappold, G.A. and H. Lehrach. 1988. A long range restriction map of the pseudoautosomal region by partial digest PFGE analysis from the telomere. *Nucleic Acids Res.* **16**: 5361.

Rasmussen, S.W. and P.B. Holm. 1978. Human meiosis II. Chromosome pairing and recombination nodules in human spermatocytes. *Carlsberg Res. Comm.* **43**: 275.

Reeders, S.T., M.H. Breuning, K.E. Davies, R.D. Nicholls, A.P. Jarman, D.R. Higgs, P.L. Pearson, and D.J. Weatherall. 1985. A highly polymorphic DNA marker linked to adult polycystic kidney disease on chromosome 16. *Nature* **317**: 542.

Reeders, S.T., M.H. Breuning, G. Corney, S.J. Jeremiah, P. Meera Khan, K.E. Davies, D.A. Hopkinson, P.L. Pearson, and D.J. Weatherall. 1986. Two genetic markers closely linked to adult polycystic kidney disease on chromosome 16. *Br. Med. J.* **292**: 851.

Reeders, S.T., T. Keith, P. Green, G.G. Germino, N.J. Barton, O.J. Lehmann, V.A. Brown, P. Phipps, J. Morgan, J.C. Bear, and P. Parfrey. 1988. Regional localization of the autosomal dominant polycystic kidney disease locus. *Genomics* **3**: 150.

Romeo, G., G. Costa, L. Catizone, G.G. Germino, D.J. Weatherall, M. Devoto, L. Roncuzzi, P. Zucchelli, T. Keith, and S.T. Reeders. 1988. A second genetic locus for autosomal dominant polycystic kidney disease. *Lancet* **II**: 8.

Rommens, J.M., M.C. Iannuzzi, B.-S. Kerem, M.L. Drumm, G. Melmer, M. Dean, R. Rozmahel, J.L. Cole, D. Kennedy, N. Hidaka, M. Zsiga, M. Buchwald, J.R. Riordan, L.-C. Tsui, and F.S. Collins. 1989. Identification of the cystic fibrosis gene: Chromosome walking and jumping. *Science* **245**: 1059.

Rouyer, F., A. de la Chapelle, M. Andersson, and J. Weissenbach. 1990. An interspersed repeated sequence specific for human subtelomeric regions. *EMBO J.* **9**: 505.

Royle, N.J., R.E. Clarkson, Z. Wong, and A.J. Jeffreys. 1988. Clustering of hypervariable minisatellites in the proterminal regions of human autosomes. *Genomics* **3**: 352.

Royle, N.J., J.A.L. Armour, M. Webb, A. Thomas, and A.J. Jeffreys. 1992. A hypervariable locus D16S309 located at the distal end of 16p. *Nucleic Acids Res.* **20**: 1164.

Saccone, S., A. de Sario, G. della Valle, and G. Bernardi. 1992. The highest gene concentrations in the human genome are in telomeric bands of metaphase chromosomes. *Proc. Natl. Acad. Sci.* **89**: 4913.

Snijdewint, F.G.M., J.J. Saris, J.G. Dauwerse, M.H. Breuning, and G.J.B. van Ommen. 1990. Probe 218EP6 (D16S246) detects RFLPs close to the locus affecting adult polycystic kidney disease (PKD1) on chromosome 16. *Nucleic Acids Res.* **18**: 3108.

Somlo, S., B. Wirth, G.G. Germino, D. Weinstat-Saslow, G.A.J. Gillespie, H. Himmelbauer, L. Steevens, P. Coucke, P. Willems, L. Bachner, E. Coto, C. Lopez-Larrea, B. Peral, J. Luis San Millan, J.J. Saris, M.H. Breuning, A.-M. Frischauf, and S.T. Reeders. 1992. Fine genetic localization of the gene for autosomal dominant polycystic kidney disease (PKD1) with respect to physically mapped markers. *Genomics* **13**: 152.

Vyas, P., M.A. Vickers, D.L. Simmons, H. Ayyub, C.F. Craddock, and D.R. Higgs.

1992. Cis-acting sequences regulating expression of the human α-globin cluster lie within constitutively open chromatin. *Cell* **69:** 781.

Wessels, J.W., P. Mollevanger, J.G. Dauwerse, F.H.M. Cluitmans, M.H. Breuning, and G.C. Beverstock. 1991. Two distinct loci on the short arm of chromosome 16 are involved in myeloid leukemia. *Blood* **77:** 1555.

Wilkie, A.O.M. and D.R. Higgs. 1992. An unusually large $(CA)_n$ repeat in the region of divergence between subtelomeric alleles of human chromosome 16p. *Genomics* **13:** 81.

Wilkie, A.O.M., J. Lamb, P.C. Harris, R.D. Finney, and D.R. Higgs. 1990a. A truncated human chromosome 16 associated with α thalassaemia is stabilized by addition of telomeric repeat $(TTAGGG)_n$. *Nature* **346:** 868.

Wilkie, A.O.M., D.R. Higgs, K.A. Rack, V.J. Buckle, N.K. Spurr, N. Fischel-Ghodsian, I. Ceccherini, W.R.A. Brown, and P.C. Harris. 1991. Stable length polymorphism of up to 260 kb at the tip of the short arm of human chromosome 16. *Cell* **64:** 595.

Wilkie, A.O.M., V.J. Buckle, P.C. Harris, J. Lamb, N.J. Barton, S.T. Reeders, R.H. Lindenbaum, R.D. Nicholls, M. Barrow, N.C. Bethlenfalvay, M.H. Hutz, J.L. Tolmie, D.J. Weatherall, and D.R. Higgs. 1990b. Clinical features and molecular analysis of the α thalassaemia/mental retardation syndromes. I. Cases due to deletions involving chromosome band 16p13.3. *Am. J. Hum. Genet.* **46:** 1112.

Wolff, E., Y. Nakamura, P. O'Connell, M. Leppert, G.M. Lathrop, J.-M. Lalouel, and R. White. 1988. Isolation and mapping of a polymorphic DNA sequence (pEKMDA2-I) on chromosome 16 [D16S83]. *Nucleic Acids Res.* **16:** 9885.

Xiao, G.H., K.H. Grzeschik, and G. Scherer. 1987. Anonymous genomic DNA sequences detecting restriction fragment length polymorphisms on human chromosomes. *Cytogenet. Cell Genet.* **46:** 721.

Zeitlin, H.C. and D.J. Weatherall. 1983. Selective expression within the human α globin gene complex following chromosome-dependent transfer into diploid mouse erythroleukaemia cells. *Mol. Biol. Med.* **1:** 489.

Index

Adenine phosphoribosyltransferase (APRT), 110
Adenomatous polyposis coli (APC) gene
  approaches for determining chromosomal localization
    allele loss, 91
    constitutional cytogenetic abnormalities, 90–91
    genetic linkage, 91
    PFGE studies, 92, 99–102
  assembly of physical map
    deletion hybrids, 92–95
    end cloning, 96–98
    irradiation hybrids, 95–96
    microcloning, 99
    PFGE, 99–101
    YAC clones, 100–101
  and FAP, 90
  as a tumor suppressor gene
ADPKD. *See* Autosomal dominant polycystic kidney disease
AHC. *See* Congenital adrenal hypoplasia
Alu sequences, 80, 84
Ankyrin repeat, 21
Antigen
  presentation, 2, 20
  processing, 20
  recognition, 71

APC gene. *See* Adenomatous polyposis coli gene
APRT. *See* Adenine phosphoribosyltransferase
ARS. *See* Autonomously replicating sites
ATP-binding cassette superfamily, 19
ATR-16. *See* α-Thalassemia/mental retardation syndrome
Autonomously replicating sites (ARS), 80, 82
Autosomal dominant polycystic kidney disease (ADPKD), 108, 110. *See also* PKD1 locus

BAT1–BAT9 genes, 15
Becker muscular dystrophy (BMD), 39–40
BMD. *See* Becker muscular dystrophy
*Bss*HII
  and molecular map of the MHC, 13–14, 17
  and physical map of the APC region of chromosome 5, 96, 98, 100
  and physical map of the MHC, 9–11

Cancer, of the colon, 90–91
cDNA sequencing, 65
Choroideremia, 36
Chromosome
  pairing, 122
  rearrangement, 123–124
  walking, 12, 37–38, 40
Chromosome 5, 89–102. *See also*
    Adenomatous polyposis coli
    gene
Chromosome 6, 2–3. *See also* Major
    histocompatibility complex
Chromosome 16, 16p13.3 region
  disease loci, 108, 124–127
  gene density, 126
  long-range restriction map
    comparison with genetic map,
      116–118
    construction of map, 114–118,
      125
    localization of 16p telomere,
      115–116
    methodology, 111–114
  polymorphism, 118–122
  primary maps
    genetic linkage map, 110–111
    hybrid map, 110
  rearrangement, 123–124
  recombination, 116–118
Chromosome 17, 2. *See also* H-2
    complex
Chronic granulomatous disease
    (CYBB), 36–39, 47–51
*Cla*I, 114–116
Complement components
  encoded by the H-2 complex, 4
  encoded by the MHC class III
    region, 2, 5
  molecular mapping, 13–16
  physical mapping, 7, 10
  tissue-specific expression, 15
Congenital adrenal hypoplasia
    (AHC), 36–37, 45–47
Conserved sequence block, 80, 83
Contigs, 38, 41–44, 46–47, 50–51
Contiguous deletion syndromes, 36,
    45, 48
Cosmids, 6, 12–17, 72–73

CpG dinucleotides
  frequency, 108, 111–114
  in Giemsa-light bands, 22
  in isochores, 108
  in the MHC class III region, 13, 15
  in the mouse TCR $C_\delta C_\alpha$ region, 80
  recognition sites for restriction
    enzymes, 8, 111–114
CpG islands. *See also* HTF islands
  associated with gene-rich regions,
    126
  in chromosome 16, 16p13.3
    region, 124
  in Giemsa-light bands, 22
  in isochores, 108
  in the MHC class III region, 13
  recognition sites for restriction
    enzymes, 96–101, 111
  in Xp21, 46, 50
CRM/grail program, 75–76
CYBB. *See* Chronic granulomatous
    disease
CYP21
  encoded by the MHC class III
    region, 5
  genetic mapping, 5
  molecular mapping, 14–16
  physical mapping, 7, 10
  tissue-specific expression, 15
Cystic fibrosis gene, 127
Cytokines. *See* Tumor necrosis factor

DCC gene, 91
Deletion hybrids, 92–95
Deletion jumping, 40
DMD. *See* Duchenne muscular
    dystrophy
DNA. *See also* Isochores
  evolutionary comparisons, 65, 69–70
  GC-rich regions, 108, 126–127
  genomic, 64
  methylation, 8–9, 13, 15, 111. *See
    also* CpG dinucleotides; CpG
    islands
  sequencing. *See* cDNA sequenc-

ing; Human genome project;
Large-scale genomic DNA sequencing
storage of genetic information, 63–65
DNA polymerase α (POLA), 36–37
Duchenne muscular dystrophy (DMD)
  clinical phenotype, 39–40
  diagnostic test for, 45
  gene for
    chromosomal location, 23, 36–37
    detection of mutations, 43–45
    diversity, 41–42
    isolation, 37–39
    physical mapping, 40–45
    structure, 40–41
  gene product. See Dystrophin
Dystrophin, 40–45

*Eag*I, 13–14, 17
End cloning, 96–98
Evolution, 69–70
Exon amplification, 38–39
Exons, 64

Familial adenomatous polyposis (FAP), 90, 92, 95
Familial Mediterranean fever, 126
FAP. See Familial adenomatous polyposis
Fibronectin type-III repeat, 21

G12–G18 genes, 15
GC-rich DNA, 108, 126–127
Gene copy number, 8
Genetic counseling, 95
Genomic DNA sequencing. See Large-scale genomic DNA sequencing
Giemsa staining, 3, 22–23
GK. See Glycerol kinase deficiency
α-Globin complex, 108, 111, 123, 126

Glycerol kinase deficiency (GK), 36–37, 45–47
Graft rejection, 3

H-2 complex, 2–4
Hardy-Weinberg equilibrium, 5
Heat shock protein (HSP70), 2, 13, 20
*Hin*dIII, 18
Histocompatibility-2 complex. See H-2 complex
HLA-A, 2, 4–6, 10–12, 18–19
HLA-B, 2, 4–6, 10–12, 18–19
HLA-C, 2, 4–6, 10–12, 18–19
HLA-D, 2, 4–6, 8–10, 16–17
HLA-E, 10–12, 18
HLA-F, 10–11, 18
HLA-G, 10–12, 18
HLA-H, 11–12
HLA-J, 11
*Hpa*II, 80
*Hpa*II tiny fragment (HTF) islands. See HTF islands
HSP70. See Heat shock protein
HTF islands, 13–17, 19, 22. See also CpG islands
Human genome
  estimated number of genes, 64, 68
  isochores, 108
  nonrandom distribution of genes, 1, 23
  sequencing, 39. See also cDNA sequencing; Human genome project; Large-scale genomic DNA sequencing
Human genome project, 65
Human leukocyte antigens (HLA). See HLA-A–HLA-J

Immune response, 2, 4, 71
γ-Interferon, 20
Introns, 64
Irradiation hybrids, 95–96
Isochores, 108, 127

Kallmann syndrome, 36, 39
*Ksp*I, 10-11, 13-14, 17

Laminin, 40
Large-scale genomic DNA sequencing
  of the $C_\delta C_\alpha$ region of the murine TCR, 74-84
  choice of regions for, 70-71
  potential benefits of, 65-70
  strategies, 67, 70, 72-74
Leucine zipper, 21
LINE. *See* Long interspersed nuclear element
Linkage disequilibrium, 5-7, 122
Long interspersed nuclear element (LINE), 78, 80, 108

Major histocompatibility complex (MHC)
  human
    biological function, 2, 71
    chromosomal location, 2
    and disease susceptibility, 6
    functions of encoded proteins, 2, 4-5, 19-21. *See also* Complement components; CYP21; Heat shock protein; HLA-A; HLA-B; HLA-C; HLA-D; HLA-E; HLA-F; HLA-G; HLA-H; HLA-J; Tumor necrosis factor
    gene density, 22
    genetic mapping, 4-6
    identification of novel genes, 12-19
    long-range restriction map, 3
    molecular mapping, 12-19
    physical mapping, 3, 6-12
    structure, 2-3
    tissue-specific expression of encoded proteins, 15-16
  murine. *See* H-2 complex
MCC gene, 101
McLeod syndrome (XK), 36-37, 48-51

Methyl adenine DNA glycosylase, 126
Methylation, of cytosine residues, 8-9. *See also* CpG dinucleotides
MHC. *See* Major histocompatibility complex
β_2-Microglobulin, 19
Microsatellite markers, 102
Mixed lymphocyte reaction, 4
*Mlu*I
  and molecular map of the MHC, 14, 17
  and physical map of the APC region of chromosome 5, 96, 98, 100
  and physical map of the MHC, 3, 12
Monosomy, 124
Multigene families, 67-69

NF1 gene. *See* von Recklinghausen neurofibromatosis gene
*Not*I, 3, 9, 12-14, 17
*Nru*I, 3, 9, 12, 14, 114-116

OED. *See* Oregon eye disease
Open reading frame (ORF), 80
Oregon eye disease (OED), 45
ORF. *See* Open reading frame
Ornithine transcarbamylase deficiency (OTC), 36-37, 48-51
OSG gene, 15, 21
OTC. *See* Ornithine transcarbamylase deficiency

PCR. *See* Polymerase chain reaction
pERT. *See* Phenol emulsion reassociation technique
PFGE. *See* Pulsed field gel electrophoresis
Phenol emulsion reassociation technique (pERT), 37
Phosphoglycolate phosphatase, 108

INDEX

PKD1 locus, 110, 124–127
POLA. *See* DNA polymerase α
Polymerase chain reaction (PCR), 13, 39, 43–45
Polymorphism, 5, 118–122
Positional cloning, 38–39
Proteasome, 20
Pseudogenes, 6, 18, 76–78, 84
Pulsed field gel electrophoresis (PFGE)
    and the APC region of chromosome 5, 92, 99–102
    and the MHC, 3, 6–12
    and 16p13.3, 112, 115
    and Xp21, 40, 44, 46, 49–51
*Pvu*I, 3, 9, 14, 17

RAG-1, 21
Rare-cutting restriction enzymes, 12–13, 16, 111–114. *See also* *Bss*HII; *Eag*I; *Ksp*I; *Not*I; *Sfi*I
RD gene, 13
Recombination, 7, 12, 116–118
Restriction enzymes. *See* *Cla*I; *Hin*dIII; *Hpa*II; *Mlu*I; *Nru*I; *Pvu*I; Rare-cutting restriction enzymes; *Rsr*II; *Sal*I
Restriction fragment length polymorphisms (RFLPs), 8
Retinitis pigmentosa form 3 (RP3), 36–37, 48–51
Retinitis pigmentosa form 6 (RP6), 46
Reverse chromosome painting, 124
RFLPs. *See* Restriction fragment length polymorphisms
RING1–RING12 genes, 17, 19–20
RP3. *See* Retinitis pigmentosa form 3
RP6. *See* Retinitis pigmentosa form 6
*Rsr*II, 113–116
Rubinstein-Taybi syndrome, 126

*Sal*I, 12
Sequenase, 73

Sequence-tagged sites (STSs), 43, 67
Serum substance (Ss), 4
*Sfi*I, 12, 40, 44
Short interspersed nuclear element (SINE), 78, 80, 108
SINE. *See* Short interspersed nuclear element
Somatic cell hybrids, 92, 95, 109–111
Ss. *See* Serum substance
Steroid 21-hydroxylase. *See* CYP21
Steroid sulfatase deficiency, 36
STSs. *See* Sequence-tagged sites
Submicroscopic terminal translocation, 124

Taq polymerase, 73
T-cell receptor (TCR)
    function, 71
    gene families, α/δ,β, and γ, 72
    sequencing of murine $C_\delta C_\alpha$ region
    C_α gene, 74–75, 77
    C_δ gene, 74–75, 77
    creation of genetic markers, 81
    evolutionary comparison with human $C_\delta C_\alpha$ region, 82–84
    J_α gene segments, 74–75, 78–79
    pseudo J_α gene segments, 76–78, 84
    repetitive sequences, 78–80
    V gene diversification, 77–78
    $V_\delta 3$ gene segment, 74–75, 78
    two types, α/β and γ/δ, 71
TCR. *See* T-cell receptor
Telomere, 115–116, 122–123, 126
Tenascin, 21
α-Thalassemia/mental retardation syndrome (ATR-16), 108, 110, 123, 126
Thymus-derived lymphocytes. *See* T lymphocytes
T lymphocytes, 2. *See also* T-cell receptor
TNF. *See* Tumor necrosis factor
Topoisomerase II sites, 80, 82

Transplantation antigens. *See* HLA-A; HLA-B; HLA-C
Trisomy, 124
Tuberous sclerosis, 126
Tumor necrosis factor (TNF), 2, 7, 13–16
Tumor rejection, 3
Tumor suppressor genes, 90–91. *See* Adenomatous polyposis coli gene; DCC gene; MCC gene

Valyl-tRNA synthetase, 21
von Recklinghausen neurofibromatosis (NF1) gene, 90–91

X chromosome. *See* Xp21
X1–X6 genes, 15
XA gene, 15
XB gene. *See* OSG gene
XK. *See* McLeod syndrome
X-linked zinc finger gene (ZFX), 36–37
Xp21
 contiguous deletion syndromes, 36
 disease gene loci. *See* Chronic granulomatous disease; Congenital adrenal hypoplasia; Duchenne muscular dystrophy; Glycerol kinase deficiency; McLeod syndrome; Oregon eye disease; Ornithine transcarbamylase deficiency; Retinitis pigmentosa form 3
 physical mapping, 35–51

Y1–Y5 genes, 17, 19–20
YAC clones. *See* Yeast artificial chromosome clones
Yeast artificial chromosome (YAC) clones
 and the APC region of chromosome 5, 100–101
 and the MHC, 11–13, 16, 18
 overlapping contiguous sets. *See* Contigs and Xp21, 38, 41–44, 46–47, 50–51

ZFX. *See* X-linked zinc finger gene
Zinc finger, 16, 21. *See also* X-linked zinc finger gene
Zoo blot analysis, 13, 37